THE BOOK OF URANUS

Also by ACS Publications

All About Astrology Series of booklets
The American Atlas, Expanded Fifth Edition (Shanks)
The American Book of Tables (Michelsen)
The American Ephemeris for the 20th Century [Noon or Midnight] 1900 to 2000, Rev. 5th Ed.
The American Ephemeris for the 21st Century [Noon or Midnight] 2001-2050, Rev. 2nd Ed.
The American Heliocentric Ephemeris 1901-2000
The American Sidereal Ephemeris 1976-2000
Asteroid Goddesses (George & Bloch)
Astro-Alchemy (Negus)
Astrological Games People Play (Ashman)
Astrological Insights into Personality (Lundsted)
Astrology for the Light Side of the Brain (Rogers-Gallagher)
Basic Astrology: A Guide for Teachers & Students (Negus)
Basic Astrology: A Workbook for Students (Negus)
The Book of Jupiter (Waram)
The Book of Neptune (Waram)
The Book of Pluto (Forrest)
The Changing Sky (Forrest)
Complete Horoscope Interpretation (Pottenger)
Cosmic Combinations (Negus)
Dial Detective (Simms)
Easy Astrology Guide (Pottenger)
Easy Tarot Guide (Masino)
Expanding Astrology's Universe (Dobyns)
Finding Our Way Through the Dark (George)
Future Signs (Simms)
Hands That Heal (Burns)
Healing with the Horoscope (Pottenger)
The Inner Sky (Forrest)
The International Atlas, Revised Fourth Edition (Shanks)
The Koch Book of Tables (Michelsen)
Midpoints (Munkasey)
New Insights into Astrology (Press)
The Night Speaks (Forrest)
The Only Way to... Learn Astrology, Vols. I-VI (March & McEvers)
    Volume I - Basic Principles
    Volume II - Math & Interpretation Techniques
    Volume III - Horoscope Analysis
    Volume IV- Learn About Tomorrow: Current Patterns
    Volume V - Learn About Relationships: Synastry Techniques
    Volume VI - Learn About Horary and Electional Astrology
Planetary Heredity (M. Gauquelin)
Planets in Solar Returns (Shea)
Planets on the Move (Dobyns/Pottenger)
Psychology of the Planets (F. Gauquelin)
Roadmap to Your Future (Ashman)
Skymates (S. & J. Forrest)
Spirit Guides: We Are Not Alone (Belhayes)
Tables of Planetary Phenomena (Michelsen)
Twelve Wings of the Eagle (Simms)
Your Magical Child (Simms)
Your Starway to Love, Second Edition (Pottenger)

The Book of Uranus

by
Joan Negus

International Standard Book Number 0-935127-27-5

Cover Design by Daryl S. Fuller

Printed in the United States of America

Published by ACS Publications
5521 Ruffin Road
San Diego, CA 92123

First Printing, March 1996

Contents

CHAPTER ONE

INTRODUCTION

Uranian Contradictions

Uranus is a complex planetary symbol that astrologically seems to contradict itself. On the one hand, it is the revolutionary planet and, on the other, is connected with humanitarian causes. Both **revolutions and humanitarian causes** elicit thoughts of strong feelings, passion. Yet Uranus is associated with **objectivity, detachment and coldness** as well. Another seeming contradiction is that Uranus represents **freedom and individuality**, but it is also associated with **equality of all** and can be important in team efforts and group cooperation.

Are these seeming conflicts irreconcilable? Perhaps not. In a horoscope it is the universe that is used to depict the individual. And just as there are few collisions in the universe, we should be able to live in relative peace and harmony here on Earth as well. The first problem we encounter in trying to make our lives harmonious is that **all of the principles connected with the planets must be accommodated in some way** and some of these principles may seem antagonistic to each other. Then we tend to concentrate on what we can easily manage and try to ignore the rest. This is when difficulties emerge with issues associated with the planets being disregarded. **The purpose of facing the issues is to bring our lives back in balance**, but this knowledge alone does not improve the situation. It is necessary to make use of this knowledge. Not only do we have to express all of the planets that comprise our horoscopes, but in some instances we have to reconcile principles connected with a single planet that seem contradictory, such as the definitions ascribed to Uranus.

We may, at certain times in our lives, attain a degree of serenity, but life is dynamic and we have to move with it. Our circumstances change and the way in which we have been expressing the principles connected with the planets may no longer be appropriate. At this point it is advantageous to be aware of the alternatives we have. **The better we comprehend the meaning of each planet, the more intelligently we can make our choices**. For example, there may be periods in our lives when we should make use of the revolutionary side of Uranus, other periods when we should remain detached, and still others when need to bring both into a situation. And the better we understand the total ramifications of the various alternatives, the better our timing will be, and the faster the balance will be regained.

Information to better grasp the meaning of the planets can come from the signs in which the planets are posited, the aspects they form with the other planets, and the houses in which they appear in the horoscope. But before we go into the specifics of astrological definitions, we can acquire some knowledge from other sources to help build a foundation of understanding.

Mythological Background

The mythology of the planets offers a wealth of data to corroborate and amplify the astrological definitions. Let us look at the mythology of Uranus. In Greek mythology Uranus was the god of the sky. Originally the sky was on earth. In fact, Uranus was married to Gaia, the earth goddess. They had many children, all of whom Uranus imprisoned within the earth — an action which is difficult to reconcile with a god whose name means sky (a place we think of as having no boundaries), or a planet we define as representing freedom. Perhaps he imprisoned his children to preserve his own independence. Gaia hated Uranus for his treatment of their children and convinced their son Cronus (Saturn in Roman mythology) to castrate his father.

There could be acceptable reasons for imprisoning one's enemies because one can feel freer if one's enemies are incarcerated. There might even be circumstances under which one should confine supposed friends. But to mistreat one's children is not easy to justify. This action leaves no doubt about the self-centeredness and lack of compassion connected with the god Uranus. So this seems to infer that in astrology, Uranus the planet, could be defined this way. We might say that Uranus at best is objective or detached, and at worst, cold, calculating, and self-absorbed.

According to his legend, Uranus was separated from the earth after his death and placed where we see the sky today. After that time there are no further stories about this god. This could perhaps indicate that once one is detached from the material plane and freedom is attained, it need not be an issue.

Keep in mind, however, that Uranus is only one planet in your horoscope and, therefore, although you need freedom and detachment in some way in your life, there are other requirements that have to be accommodated in some way as well.

Astronomical Notes

The astronomy of Uranus can also shed light on the astrological meaning of the planet. The "backward" rotation of Uranus can be correlated with the concept of eccentricity associated with that planet. The oddness of Uranus is further compounded by the fact that, astronomically, its equator is inclined 98 degrees with its orbit, and the planet gives the impression of rolling along on its side.

In 1977, however, another facet of Uranus was seen for the first time. Five rings were observed around the planet. Further surveillance brought the total to nine, and photographs later sent back from Uranus by Voyager confirmed these findings. So Uranus has rings. The revolutionary, nonconforming qualities ascribed to Uranus work too well in a chart to discard. But the message of the rings would indicate that there may be a limit to the amount of independence you can or should have. Possibly, by recognizing and defining your limitations, you can feel freer. Or, it can mean that if you choose to stretch your boundaries, you must first be aware of what the parameters are. Then you can reach beyond them.

Historical Notes

Conditions in the physical world can also contribute to our understanding of the planets. The three planets that were discovered in modern times (Uranus, Neptune and Pluto), seemed to have appeared in the sky at times when the definitions ascribed to them were appropriate. In the case of Uranus, it was discovered in 1781, during the period of the French and American revolutions. This was also during the period of the Industrial Revolution. So **Uranus is connected with modern technology as well as political revolutions**. And each time an innovative idea is projected into the world, Uranus will probably be in the picture. If the innovative ideas are being expressed in the individual's personal life, then Uranus should be prominent at that time in the native's horoscope. But if the innovation has an impact on the greater world, it should stand out in mundane charts.

Largely because Uranus is associated with all forms of revolution, it was assigned co-rulership of Aquarius, the sign that we think of as revolutionary. By looking at the traits associated with the sign of Aquarius we may also get insights into the planet Uranus.

Uranus as Ruler of Aquarius

Aquarius is an air sign. Air signs are **mental and objective**. Aquarius is considered the most **intellectual** of the air signs. It is therefore reasonable that Uranus, the higher octave of the planet Mercury, should be the co-ruler of Aquarius. The objectivity associated with the sign can make Aquarians good mediators. They have the ability to see all sides of a situation when they are not personally involved, and thus can be fair and just in making decisions. There is, however, an **aloofness** connected with Aquarians and the true Aquarian cannot be reached through the emotions. Compassion is not a key word for this sign. If the

Aquarian feels personally threatened, the impartiality and detachment can turn into absolute coldness—as evidenced in the myth of Uranus.

Aquarians can start revolutions and fight for the cause of justice. But it will be the intellect rather than emotions that motivates them. It is usually the political or social cause, not the people, that gets them moving.

We can apply these ideas to the planet Uranus, since this planet feels comfortable in the sign of Aquarius. So you may be predominantly conforming, or mild-mannered, or opinionated, but the planet Uranus will show you where and how you can be **individualistic, revolutionary or objective**.

To continue with the kinship of Aquarius and Uranus, Aquarius is considered an original and creative sign, qualities we also associate with the planet Uranus. Thus, Uranus in your horoscope will give you information about your **originality and creativity**. Since **it is in a sign for about seven years**, there may be certain qualities that are generational in regard to Uranian matters such as originality, creativity, freedom and revolutionary tendencies. You may, therefore, share certain ideas and attitudes on these matters with people born during the seven-year period when Uranus was in the same sign as yours. But Uranus becomes more personalized when we add the houses in which it is placed in the horoscope and the aspects it forms with other planets and points in the chart.

In this book we will examine Uranus from all of these perspectives. We will discuss the many ways in which you can experience the planet, with an eye on learning to use it most effectively in your life. You will be given the ingredients and it will be up to you to create the recipes that will be the most appealing to you.

CHAPTER 2

URANUS IN SIGNS

Uranus completes a cycle in 84 years and it is **in each sign for about seven years**. Because of its slow velocity, it not only stays in each sign for a long period of time, but it also forms long lasting aspects, particularly with the other outermost planets. Therefore mundane astrologers have used Uranus to explain world events. As a result of that, many believe Uranus transcends the individual. For this reason it is sometimes called a "transpersonal" planet.

It is true that the sign placement of Uranus at any given time can describe the **attitude of the world toward change and matters of individuality and freedom** during that period. The types of mundane events we might look for are revolutions, or situations in which freedom or individuality is in the foreground. For example, it is interesting to note that United States came into being as the result of the American Revolution and, **in the natal chart of the United States, Uranus is in the sign of Gemini. When transiting Uranus returned to this position 84 years later, the American Civil War took place, and 84 years after that was World War II**. Not only was there revolutionary action during these times, but each war had a great deal of **rhetoric** connected with it, and **important documents** were an integral part of the total picture.

Modern technology is also associated with Uranus and we can form a better understanding of the planet by looking at what technology was developed during its stay in each sign. As pointed out in Chapter One, the discovery of Uranus coincided not only with the French and American Revolutions, but also with the onset of the Industrial Revolution. **Inventiveness and creativity** are prerequisites for technological development and thus are easily added to the definition of Uranus. So looking at the creative process behind inventions can help us to more fully grasp the meaning of Uranus in each sign.

The sign position of Uranus at the time of your birth, however, not only describes conditions in the world at that time. It can also provide personal information. Mundane events occurring during a particular period show the operation of the mass consciousness, which can be associated especially with the three transpersonal planets (Uranus, Neptune and Pluto). The sign position of each of these planets can be used to explain happenings or movements in history that are appropriate to their astrological definitions.

There is **little free will in mass consciousness**. As a group, human beings may be swept along by the tide of public opinion, but there is much more latitude in personal behavior. In your private life there are always choices to be made, no matter what is currently happening in society. The manner in which you operate is shown in the birthchart. Since the birthchart is a transit chart of a moment in time, at a given location, it connects the essence of that moment to individuals who are born in that place at that time. Therefore, it not only explains world events, it also describes the manner in which the native will behave throughout his or her lifetime. So if you look at the sign in which Uranus appears in your horoscope you will better understand Uranian issues that were prevalent when you were born. But this in turn provides **information about your own concept of freedom and individuality** and how, in general, you will view and express these qualities. It can also give **clues to your inventiveness**.

You will probably notice that **you have much in common in terms of attitude and perception of Uranian issues with most other people born during the seven-year period in which Uranus is in the same sign as your own**. You will not, however, be a clone of everyone born with Uranus in that sign. As the planet is combined with other factors in the horoscope, your ideas and sentiment toward matters of freedom and individuality may be modified or supported. When you look at the house in which Uranus is placed, and the aspects it forms with other planets and points in the horoscope, the definition of this planet becomes more specific and personalized. The aspects of other planets to Uranus at particular times will also have an impact on world events and, where appropriate, will be described in the chapters on aspects. Yet, interpreting Uranus solely in terms of the sign in which it is posited is still valuable.

As we discuss Uranus in the signs we will examine world events, and then apply the implications to a person born at that time. This may not only broaden your view of history, but also help you to learn something new about yourself. And the better you grasp each part of your total make-up, the more likely you will be able to build toward your sense of wholeness.

During the twentieth century Uranus will have moved through the entire zodiac and the signs of Sagittarius, Capricorn and Aquarius twice. It was at 10° of Sagittarius on January 1, 1900 and will be at 14° of Aquarius on December 31, 1999. As we discuss Uranus in signs, the years in which it was moving through each sign will be mentioned. There will be some years in which Uranus will move between two signs, so those years will appear under both signs. This is because Uranus was or will be moving forward from 29° of one sign to 0° of the next and/or going retrograde from 0° back into the preceding sign. If your Uranus is at the very beginning or end of a sign you will probably find some of the definitions of both your sign and the one preceding or following will be applicable to you. For example, if you have Uranus at 29° of Leo, it is so close to 0° Virgo that characteristics of each of the two signs might be evident in Uranian type of activities or your attitude toward Uranian issues. You might dramatically announce your revolutionary activities (Leo), but plan out every detail before you take action (Virgo).

The periods of transition are particularly significant in terms of mundane events because the shift in emphasis will be evident in the public's attitude toward change and the nature of Uranian type of events. The change may not be obvious immediately. You may not see a drastic change in attitude the moment that Uranus moves into 0° 0' of the next sign. The transition starts as the planet nears the last few degrees of a sign and becomes increasingly clear as it moves through the first few degrees of the next sign. Therefore, particular attention will be paid to appropriate events occurring during these periods.

As each sign is covered, we are most interested in what the particular sign placement could mean for the individual, but examples of mundane events will be mentioned to help clarify the meaning.[1] If your Uranus is in a particular sign, you might find it valuable to probe more deeply into the period of history that includes your birth year.

---

1 The book *The Almanac of American History* by Arthur M. Schlesinger, Jr., New York: Bramhall House, 1983 is the main source for the mundane information.

CHAPTER 3

URANUS IN SIGNS: SAGITTARIUS, CAPRICORN, AQUARIUS

URANUS IN SAGITTARIUS (1900-1904) (1981-1988)

Uranus was already in Sagittarius as we entered the year 1900. In 1899, the US asked for an "**open door**" policy with China by which all countries would be treated equally in terms of trade with China. In 1900 Russia, Germany, France, England, Italy and Japan agreed. Just the words "open door" implies the Uranian idea of freedom and the Sagittarian connection is inferred because fairness was an issue and foreign countries were involved.

Another example of Uranus in Sagittarius occurred in China with the **Boxer Rebellion**. The revolution was to protest foreign domination. The revolutionaries called themselves "righteous harmony fists." Property owned by foreign powers was destroyed and, after the rebellion, compensation was paid to the countries involved. In 1908 (when Uranus was firmly entrenched in Capricorn), the United States returned some of its settlement to China for practical purposes of education.

The late 19th and early 20th centuries also coincided with the time of the **robber barons** of industry. These men were expanding (a Sagittarian keyword) their corporations and forming monopolies in defiance of the Sherman Anti-trust Law. They considered themselves above the law. By February 1902, President Theodore Roosevelt was waging war on the monopolies in order that the government might exercise more control over them. His first step was to bring suit against the Northern Securities Company. This case was not settled until two years later—the year in which Uranus moved into Capricorn.

Exploitation of their employees was also part of the lack of concern of these captains of industry. Although deplorable conditions existed long before 1904, it was not until that year with Uranus moving into the earth sign of Capricorn, that something concrete was done about it. Textile workers in Fall River, Massachusetts waged a long and bitter strike. This brought to light the mistreatment of employees, including children, which led to a Child Labor Committee being formed in Congress and, eventually, to the protection of all children through **Child Labor Laws**.

In the area of modern technology, the first cable was laid between San Francisco and the Phillipines in 1902 and the **Wright brothers made their first flight in 1903**. Both of these occurrences are clearly Sagittarian examples. The cable stretched beyond our domestic borders and the airplane is not only connected with distant travel, but this invention is also responsible for bringing the world closer together in terms of time.

During the next period of Uranus moving through Sagittarius (1981-88), technological developments were centered on **space travel and computers** — emphasizing the revolutionary broadening of our horizons in true Sagittarian fashion. The idea of "**trust busting**" came before the public again with the suit brought against the American Telephone and Telegraph Company, after which AT&T had to relinquish rights to local phone companies. Revolutions included the fall of the government in Haiti in 1986 and US intervention in Grenada. Human rights were important in both cases and correlate with Sagittarian idealism.

If we look at the overview of the mundane affairs mentioned above, we find two contradictory themes. On the one hand, we have the robber barons expanding and developing without thought of the welfare of others — making their own laws or seeming to feel above the law. On the other hand, we see the emphasis on fairness, justice and benefits for humankind in the improvement of conditions of workers and the motivation behind the revolutions.

Both these themes can be evident in individuals with Uranus in Sagittarius as well. There may be a disregard for others if one's own freedom is in question or if people are interfering with one's expansion or development. But there is also an idealistic quality that is prevalent when others are being mistreated and one is not personally involved.

If you have Uranus in Sagittarius, you may be a visionary or an idealist in terms of your philosophy regarding freedom and individualism, but actions do not always follow or accompany the words. Moral consciousness may be an issue in your evaluation of world conditions, but in your own life, your urge for progress may override your own conscience. You could espouse ideas of justice and equality for all and discuss changing world conditions. But when you are exhilarated by challenge, and you are using your ingenuity in broadening your own horizons, the matter of right or wrong might not be considered. So, although your enthusiasm may cause you to ignore your righteous philosophy, the philosophy is still there. And once you are conscious of your actions not always reflecting your belief system, you can correct the situation. You should still try to be innovative and move ahead, but keep your philosophy in mind as you are personally trying to develop. This should not only keep you within acceptable boundaries, but by applying your moral viewpoint to yourself as well as to the world around you, you cannot help but improve both your position and that of those with whom you come into contact.

The theoretical side of freedom and justice may seem more obvious than tangible results stemming from actions taken. It is easier to philosophize than it is to work toward specific, concrete ends. This is clear in terms of the mundane data given above. The deplorable labor conditions in factories were exposed while Uranus was in Sagittarius. And the anti-trust suit against the Northern Securities company was started during that period as well. But it was not until Uranus moved into Capricorn that labor laws were passed and the government won its suit.

In terms of creativity, if you have Uranus in Sagittarius, you will probably find anything you produce that fits this category will have as its purpose the betterment of humanity or the broadening of personal horizons. This latter can include bringing the world closer together (as illustrated by the invention of the airplane) while you are expanding yourself. So if you have Uranus in Sagittarius and are having difficulty being creative, focus on your motivation. This might help to get you started.

URANUS IN CAPRICORN (1904-1912) (1988-1996)

Although Uranus in Capricorn may not be as free or flexible as it is in Sagittarius, its impact is clearer and more tangible, not only in the examples given above, but also in a number of other mundane events. The devastating San Francisco **earthquakes** of 1906 and 1989 both had Uranus in Capricorn, and at the time of each quake Uranus was also conjunct the San Francisco Midheaven (MC). But San Francisco is not the only place where earthquakes have occurred. There have been an inordinate number of major earthquakes all around the world since Uranus moved into Capricorn in 1988.

Another type of Uranus in Capricorn activity evident during the early Uranus in Capricorn period (1904-1912) was that the unions gained power through protection of the labor force. Strikes by workers—in protest of conditions—were becoming more frequent because of the **strength of the unions**. This fits well with the concept of authority associated with Capricorn. And with Uranus in Capricorn you would expect that it is easier to rebel if you have a group representing some form of authority supporting you and giving you a sense of security.

The idea of authority was also obvious in the American political arena. Theodore Roosevelt, in his first annual message to Congress, announced that in accordance with the **Monroe Doctrine**, the United States would deal with any problems which foreign countries had with countries in the Western Hemisphere. In his words, "Chronic wrongdoing...may force the United States...to the exercise of international police power." In 1905, he was called upon to fulfill his promise when the US government began to supervise the Dominican Republic's payment of national and international debts.

Again **in 1906 the United States assumed the role of authority in another country in the western hemisphere** when the Cuban president asked the United States for assistance in quelling a rebellion. William Howard Taft was sent to head a provisional government and the United States dominated Cuba for three years. It is interesting to note that **in the late 1980s, relations with Central American countries and the degree of United States intervention there, came again in the foreground of American politics as Uranus repeats its transit through Capricorn**.

There are other mundane examples of Uranus in Capricorn that could be mentioned, but the flavor of this placement should be clear from the illustrations already given. In fact, the earthquakes alone project an image that epitomizes the essence of Uranus in Capricorn. The earth literally split apart suddenly, indicating that the revolutionary qualities associated with Uranus can produce visible results in the sign of Capricorn. However, although the occurrence itself was abrupt, the conditions of the earth had to build over many, many years to a point of the eruption. Therefore, we lived with the knowledge that an earthquake could take place at any time, but the exact moment was unpredictable, and the devastating results a shock.

Applied to the individual, **if you have Uranus in Capricorn**, you probably live with the knowledge that if you start a revolution you can make a difference. The question will be, is it worth your time and effort? Therefore, your spontaneity is tempered by the need to evaluate and plan. You will probably rehearse and memorize your actions before you "suddenly" move. And, when you do take action, it will probably be noticed.

Not only are your "spontaneous" acts most likely planned, but the route you take to reach your goal is probably not original. You will find it easier to rebel if there is precedence to follow, or if you have the support of some segment of society. It may not be difficult to run the world with Uranus in Capricorn, but, **before you start the revolution, you want to know that you have the capability**

of overthrowing the present regime. Secondly, you want to make sure that you have subjects who will follow you when you do take over.

In other words, with Uranus in Capricorn, drastic action will rarely be taken without investigation and having precedents to refer to, which facilitates you making decisions in terms of revolutionary action. While the motivation for rebelling with Uranus in Sagittarius is usually fairness and justice, the reasons for such movement with Uranus in Capricorn are **law and order, and paternalism**. Roosevelt's announcement that we will police our own indicates the stern father approach, and the unions watching over and protecting the workers reiterates this idea.

Both these examples seem to have **law and order** as a factor. Roosevelt's attitude was to keep the Dominican Republic and Cuba in line; the unions wanted laws passed to protect the workers. But there is another example which is even more impressive. In 1910, as Uranus was coming to the end of Capricorn, a revolution broke out in Mexico. The reason for the uprising was that Mexican Dictator Porfirio Diaz and his friends were becoming rich by robbing the poor. When Taft investigated, he ignored the mistreatment of the poor. Instead he praised Diaz for his ability to maintain "law and order." He also stated that backing up the dictator was necessary to protect the large amount of capital that the United States had invested in the country. Support of the ruthless government of Mexico continued until 1914 when Uranus was in the sign of Aquarius.

URANUS IN AQUARIUS
(1912-1920) (1995-2003)

The shift in attitude toward the Mexican government that took place in 1914 concerned matters of principle rather than finance. The incident that triggered the change was the arrest of several sailors and officers from the American barge *Dolphin* in Tampico. The men were released but Admiral Henry Mayo demanded an apology and a "salute to the American flag." The Mexican dictator,

Victoriano Huerta, refused and broke off diplomatic relations with the United States.

European countries with vested financial interests in Mexico urged US President Woodrow Wilson to recognize Huerta, stating that he was strong enough to keep the status quo in this country where rebels were again threatening to challenge the government. Instead, the usually peace-oriented Wilson, asked and received permission from Congress to send troops to control the city of Veracruz. The reason given for this action was that Germany was sending arms to Huerta through this port. The incident might have triggered a war had not Argentina, Brazil and Chile entered the picture as mediators. By stopping the shipment of supplies, however, Huerta's power diminished and eventually he resigned and left the country.

This series of events points up two important differences between Uranus in Capricorn and Uranus in Aquarius. The first is that financial and other material considerations that are foremost in the minds of people with Uranus in Capricorn, fade into the background when freedom and individuality are threatened. With Uranus in Aquarius the words "personal dignity" or "personal integrity" are mentioned as being an essential ingredient in one's personal make-up and a prime reason to rebel.

The mediation that took place illustrates the second difference. The dictatorial, paternalistic approach of Capricorn is replaced by the **objectivity of Aquarius**. There is no longer the necessity of being the authority figure. We are dealing with the sign of equality, where opinions and alternatives can be discussed. With Uranus in Aquarius you will believe that you have an open mind when it comes to change and freedom, and you are capable of adjusting. You just may not do it as quickly as you think you can (or do). But at least compromise is possible. And what others have to say regarding Uranian matters can eventually be accepted if the arguments make sense and humanitarian reasons are offered.

Capricorn is a black-and-white sign. Things are right or wrong, and tangible actions must be taken to reinforce what is right and eliminate what is wrong. There is usually "one way" to resolve issues and this is dependent upon who is in charge.

With Uranus in Aquarius, the approach to Uranian issues is more abstract. "Principles" outweigh specifics. There is an awareness of the possibility of choice and recognition that there are alternative courses of action. Decisions are based on intellect, not on feelings or concrete facts. And if you try to influence a Uranus in Aquarius person with a sad story or fifty practical reasons for starting a revolution, you will probably experience a cool response. When emotions or too much practicality are brought into the picture, the Uranus in Aquarius person has the ability to detach from the situation and withdraw. **The best way to reach the Uranus in Aquarius person is to be impersonal and objective**.

An excellent example of the difference between the two signs was evidenced at the time of the passage of Uranus from Capricorn to Aquarius in 1912. That was the year in which Woodrow Wilson was elected president in a landslide victory over both Theodore Roosevelt and William Howard Taft. The events that took place in each of their terms of office which were mentioned earlier illustrate clearly the mood of the country during those times, and a look at the general character of the three men reaffirms the information —but most particularly the contrast between Teddy Roosevelt and Woodrow Wilson, as Taft worked under Roosevelt and really rose from his shadow.

Roosevelt's motto was "speak softly but carry a big stick." This evokes the image of the **stern father** whose authority is felt without it being spoken. It was time to make concrete changes materialize with Uranus in Capricorn. Therefore, a man with the character of Roosevelt would appeal to the populace during that period.

When Uranus moved into Aquarius, different types of changes were desired. The people wanted to work toward their hopes and wishes — to aspire more highly — and perhaps improve conditions for all of mankind. Using physical force was no longer enough. **The mind** needed to be exercised as well. Therefore, it is not surprising that Woodrow Wilson, **a former professor** and president of Princeton University, would be elected President of the United States. Nor is it surprising that he would be considered a visionary and noted for his intellect.

Although these years of Uranus in Aquarius (1912-1920) coincided with using the mind to institute change and to work toward higher ideals, it also included World War I (1914-1918), thus demonstrating that the intellect alone may not be enough to alter conditions. In fact, it suggests that if the mind cannot make the difference, any feasible means can be used. With Uranus in the sign of its rulership, we have a principle associated with both the planet and the sign of Aquarius, strongly expressed through World War I. With this air combination, neither feelings nor practicality play an essential role in actions taken to bring about change or freedom. The end justifies the means—no matter how unorthodox or, in this case, how bloody—it may be.

This example can add more information to our interpretation of Uranus in Aquarius for the individual. **If you have this placement, you may view alternatives for change first through the intellect**. Yet, you may be willing to defend your own right to freedom and equality in any way that is expedient. The same holds true if you become involved in a humanitarian cause that does not involve you personally. You would probably join such a cause for idealistic reasons. There need be no close connection between you and others defending or affected by the project. In fact, **you can probably start and continue revolutionary activities in an impersonal, detached manner**. It is easier to embrace principles than people, and sentimentality is probably not a consideration in Uranian matters.

CHAPTER 4

URANUS IN SIGNS: PISCES, ARIES, TAURUS

URANUS IN PISCES (1919-1928)

Uranus in Pisces presents a marked contrast to Uranus in Aquarius. The objectivity of the air sign is gone. **Feelings run strong** and the victim-victimizer theme comes to the foreground. The **Versailles Peace Treaty** (1919) illustrates this very well. The Germans had to admit guilt and give up rich territory. There seemed to be no mercy, only severe punishment for wrongdoing. A second Piscean quality that was evident in this document was its vague language which Hitler later was able to distort to his own advantage (when Uranus moved into Aries).

Another excellent example of Uranus in Pisces came as the result of the passage of the **18th Amendment banning alcohol in the United States**. Ratification of an Amendment to the Constitution is not a Uranian matter. It did, however, concern the consumption of alcohol, which is associated with the sign of Pisces. But more significant was the rebellion and the crime that it led to.

The prohibition of alcohol led to bootlegging. It was the era of speakeasies and gangsters. Defiance of the establishment was a way of life for these people. As with any revolutionary group, they had their own code of ethics, but their tenets were not based on justice and progress (Sagittarius). Nor did they care about precedence and facts (Capricorn). Nor were they objective, fair and intellectually sound (Aquarius). They were founded in feeling (Pisces). The prominence of Pisces is further reiterated in the **victim-victimizer theme** which was so frequently part of the scenarios in the interplay of the mobs in the gangster era.

Ideology is also associated with the sign of Pisces. The **American Communist Party** was founded in 1919, as Uranus was moving between Pisces and Aquarius. But it was not so much the formation of the group that epitomized Uranus in Pisces. Rather, as with the above example, it was the aftermath that was most significant. Although there was no real threat of war, the attitude of the country was hysteria in terms of this new movement. In January of 1920, United States Attorney General A. Mitchell Palmer coined the phrase "Red Menace," to describe this group, and triggered an uncontrollable emotional response in the country. **In one night he had 4000 people in 33 cities arrested as Communists**. Many of these people were not even Party members and none of them seemed to be planning a violent overthrow of the government. Yet the mood that Palmer evoked swept across the nation and led to other arrests that were also without real cause. There were **no facts nor logic connected with these incidents**—just unleashed emotions. These actions also point up the illusionary and delusionary qualities associated with Pisces.

Clearly, **if you have Uranus in Pisces** and you could easily become involved in some kind of cause, you will most likely select one that is connected to victimization or ideology. It is your emotions that draw you in, and once you are committed, hard, cold facts will probably not influence you. Your concept of the other side—the enemy—may be somewhat vague and defy concrete description. But you feel the "menace." You know it is there.

Often with Uranus in Pisces there is a strong undercurrent in regard to the process of change. Much **more is brewing under the surface than comes out into the world**. And, of what emerges, not all is announced or even recognized, whether it concerns a country or a person. The end result just seems to appear from nowhere, and one may wonder how the changes took place. Who can explain in detail how the gangsters came into power, or precisely what caused the mass hysteria over the Communist Party? It is difficult to ascertain because **emotions take over** and, therefore, rational explanations are irrelevant. There is also an **air of secrecy** that can surround matters associated with the sign of Pisces, so you can also be quiet about your nonconformity. Besides, you know that "right is on your side" and that's what counts.

You can convince yourself that your cause is the only universally right one, no matter what the tangible evidence shows. If doubt creeps in, you can, at least for a time, create an illusion and mask the flaws. In terms of **your personal quest for individuality and freedom, you can justify anything you do by ascribing some moral or spiritual reason** to it. I remember someone with this placement explaining that she took some bizarre, seemingly self-centered action because she was "getting in tune with the Universe." There was undoubtedly some of the same tone in the air among the mobsters, the American Communist Party, and even the Attorney General of the United States to account for the Uranian actions they took. When people disagree with you, you are surprised. You may become a fanatic in trying to enlighten them. Or you could simply ignore them because you have right on your side, and "they will get theirs!"

But having a righteous attitude such as yours has a positive side as well. When you are championing the cause of the downtrodden or the victimized, you can be an inspiration to those you are trying to help. You will also be a force to be reckoned with because your emotional fervor will give you the impetus to battle the enemy and ultimately wear them down no matter how strong they appear.

When you have Uranus in Pisces, the key to using it most auspiciously is to keep in mind that being of service is an essential ingredient in your taking revolutionary actions. You might become more highly evolved in the process, and even possibly reap rewards, but helping others to express individuality or to experience personal freedom should be an integral part of your motivation. When you consider the needs of others, "right" will be on your side, and your faith in your cause or direction can give you the boundless energy to accomplish seemingly impossible tasks in an apparently effortless manner.

URANUS IN ARIES
(1927-1935)

In contrast to Uranus in Pisces, **Uranus in Aries is much more direct and overt**. The keyword for this placement is "action," and little thought may be given to the consequences thereof. Sudden changes are made or revolutions precipitated not because God says you should (Pisces), nor society demands it (Capricorn), nor for reasons of fairness, equality and justice (Sagittarius and Aquarius), but rather because "I" say it's right. There is the attitude of "what's in it for me," or "if it's right for me, it's right for everyone." This is one way to explain why Hitler was able to come to power when Uranus was in Aries in contrast to why Teddy Roosevelt appealed to the people when Uranus was in Capricorn.

Teddy Roosevelt was the strong defender of a nation. He stood for the rights of the establishment. **Hitler**, on the other hand, represented the aggressive energy of the person who will stand up for his supposed personal rights and overcome and dominate the victimizer. This was the spirit he had to arouse in the people in order to get their support. He capitalized on the alleged persecution of the German nation that took place while Uranus was in Pisces. And when Uranus was in Aries, he led the fight to conquer those who had oppressed Germany.

Fortunately, not everyone with Uranus in Aries is an Adolph Hitler. We would have to look far beyond one planet to describe the world attitude that made his deplorable actions possible. Besides, we do have another world leader who emerged on the scene while Uranus was in Aries and in many ways embodied the more positive side of the sign. This man was **Franklin Delano Roosevelt** who implemented unusual policies and revolutionary ideas to combat the "Great Depression." He too moved ahead because something had to be done, and he appealed to the American public's need to improve its situation. The voters elected FDR to the presidency not for idealistic reasons, but rather because each believed that this man, as a powerful man, would do something for "me."

Roosevelt's predecessor, Herbert Hoover, was elected in 1928 with the slogan "A chicken in every pot, a car in every garage." It was in January 1928 that Uranus moved into Aries to stay for the next few years. The slogan epitomizes the self-interest that is associated with the sign of Aries and, was reflected in the mood of the country that led to Hoover's election. The air of optimism at that time cannot be explained solely by Uranus in Aries. It was compounded by Jupiter forming a conjunction with Uranus (see Chapter 13, page 128) during that period. But, as already stated, a complete picture of world conditions can only be created by an examination of all the planets at the time in question.

However, **the optimistic, risk-taking quality ascribed to this combination, especially in the sign of Aries, can help to explain the attitude and conditions that led to the depression**. You need look no further for astrological reasons or excuses for the marginal buying that contributed to the Stock Market crash in 1929. But Uranus alone, in Aries, can justify why the populace was eager to follow the unconventional ideas of Franklin Roosevelt so willingly. Hoover had failed. It was time to try something new to better the circumstances of the individual. So we can add to our definition of Uranus in Aries the **adventurous, pioneering spirit that those who have this placement can exhibit in their revolutionary activities and their search for freedom and individuality.**

URANUS IN TAURUS
(1934-1942)

As Uranus moves into Taurus, **planning and practical results are requirements** of actions taken in the name of freedom or individuality. The **need for tangible reasons and visible rewards for revolutionary acts is a quality shared by Uranus in all of the earth signs**, as has been shown with Uranus in Capricorn and will be seen again with Uranus in Virgo. The specific goals, however, are different, as demonstrated by comparing the twentieth century U.S. presidents who were in office during the period when Uranus was in Capricorn and in Taurus.

Theodore Roosevelt was president during the era of Uranus in Capricorn and his cousin, **Franklin, when Uranus was in Taurus**. Both men were Uranian types in that they were responsible for dramatic changes in the country, and in their own ways were revolutionaries. But their methods, and the images they projected, were not the same.

Teddy Roosevelt had the qualities of a stern father figure and demonstrated the might of government. Many of the changes for which he was responsible, although they ultimately were drastic, did not seem to happen suddenly. They evolved naturally from circumstances that were begun with Uranus in Sagittarius. There were really few surprises. FDR, on the other hand, came to power while Uranus was in Aries, so that the changes he instituted at the beginning of his term in office were revolutionary and contributed to his image. His ideas and actions were radical. He appeared to the public as a pioneer rather than a father figure. Then Uranus moved into Taurus and it was time for his ideas and actions to bear fruit.

Since he was reelected three times and remained in office during that entire period that Uranus was in Taurus (and beyond), he must have succeeded in his efforts. And there is physical evidence that he did. Among the main issues when he took office were difficult economic conditions in the country due to the depression. Therefore, **many of the changes he initiated had to do with finances and that, of course, fits well with the sign of Taurus**.

Two revolutionary changes he instituted were the **establishment of the Securities Exchange Commission** to oversee the stock exchange, and the **creation of Social Security** which guaranteed pensions (thus offering material security) to those retiring at age 65. Another creative innovation which was attributed to FDR and illustrated another facet of the practical sign Taurus was the establishment of the **Works Progress Administration** which employed people to build roads, bridges and public buildings.

Adolph Hitler also provided an insight into Uranus in Taurus. As with FDR, Hitler came into power during the time that Uranus was in Aries, and was seen as a pioneer in his way, too. As was stated earlier, he also exemplifies the aggressive, self-serving qualities we identify with the sign of Aries. However, it was not until **1935 that the Saar, which had been taken away from Germany after World War I, was returned to Germany—a tangible reward for Hitler's revolutionary actions.**

The sign of **Taurus shares with Aries the idea of self-interest, but replaces Aries impatience and lack of planning with plodding practicality**. Acts of war were perpetrated while Uranus was in Aries, but it was not until Uranus moved into Taurus that it seemed to be fully acknowledged or even needed to be labeled. The United States did not officially enter the war until Uranus was in the last few degrees of Taurus, reinforcing the slow-moving quality associated with the sign.

If you have Uranus in Taurus, you have probably noticed that you do not usually take drastic action without consideration of alternatives and possible results. You may look at others who are more spontaneous either with envy because you think you would like to be that way, or might view them with amazement, wondering how they could be so foolhardy. It is not that you are incapable of revolutionary action. You just want to make sure that it is worth your while. Once you have decided and embark on a course of action, you will usually follow through. After you have determined your goal, you can move unrelentingly toward it. When your mind is made up, you will rarely be influenced by the opinion of others.

Although you would expect the originality of Uranus to be squelched in Taurus, it is still possible to use ingenuity and take unorthodox routes to attain your goals. It is only the end result that must be practical and tangible. The examples given above attest to that. FDR certainly used **unusual ways to produce very down-to-earth results**. So if you have Uranus in Taurus in your natal chart you should first ascertain your practical goals. Then consider your course of action. You may discover that you have original ideas and that you are freer and more creative than you ever thought you were.

CHAPTER 5

URANUS IN SIGNS: GEMINI, CANCER, LEO

URANUS IN GEMINI (1941-1949)

As mentioned in the beginning of this book, **Uranus was in Gemini during the American Revolution, the Civil War in the United States and World War II. The Civil War and World War II**, as also previously indicated, occurred when Uranus returned to its natal position in the U.S. horoscope. It is not strange that a revolution would occur at the time of a Uranus return, but how can we reconcile Uranus being in the sign of Gemini?

Since Gemini is an air sign, we would expect some form of **communication** to lead to dramatic Uranian changes and at least **one important document** stood out during each war to support this premise. **The Declaration of Independence**, signed in 1776, set the stage for the United States becoming an independent nation. The **Emancipation Proclamation** presented by Lincoln to his cabinet in 1862 had a great impact on the Civil War, and ultimately resulted in the freeing of the slaves. And the **Charter of the United Nations** created in 1945 will hopefully eventually lead to the freedom of all nations and worldwide cooperation. There were also outstanding **orators** connected with each war: **Patrick Henry** during the Revolutionary War; **Abraham Lincoln** during the Civil War; **Franklin Roosevelt, Winston Churchill and Adolph Hitler** during World War II.

As we focus specifically on the period in the twentieth century in which Uranus was in Gemini, another Gemini quality becomes evident in connection with revolutionary matters. Communications played an important role in our attitude toward World War II. The **movies and books of that era were filled with messages to evoke patriotism**. It was as though we had to be convinced that the actions of our country were the right ones in order to stay on that path. And the propaganda continued throughout the war.

Another Uranus in Gemini characteristic that was obvious in mundane events of that period was speed. When Uranus was in Taurus, we seemed to be heading unrelentingly in the direction of a worldwide conflagration, but the movement was slow and steady. Once Uranus transited into Gemini the pace accelerated. There seemed to be more skirmishes and isolated battles. This was particularly clear as the war spread to the area of the Pacific Ocean. Battles took place on islands and thus there were **interruptions in the action** as the troops moved from place to place. In Europe, however, (where the war began with Uranus in Taurus) soldiers traveled uninterruptedly across land.

So **if you have Uranus in Gemini**, you will feel quite comfortable with the spontaneity associated with Uranus. You might not stay with a revolution or cause for an extended period of time, but you can react and become involved quickly. Then, of course, you can drop these activities just as suddenly. Because you want to remain as unencumbered as possible, you will probably not get too close to the people involved or too deeply attached to the activities. In this way you can easily flit to some new cause. If you are too interdependent with the other participants or at the hub of activity, it is more difficult to remove yourself and you could feel trapped.

You really are not looking for deep-seated commitment in terms of revolutionary activities. If you become interested in a cause, **you probably want to make your contribution quickly and go on to something new before you become bored**. You could be chosen as **spokesperson** for the group or do writing in connection with the project. Your **objectivity** and ability to reach people on their own level would help you explain issues clearly.

When you are dealing with matters of change, freedom and originality in your own life, you could find that **talking about possibilities before you take action is helpful** and perhaps even essential. Then once you think you have made up your mind you might make quick, but small changes. You could ultimately make drastic changes, but you should do this in stages. In fact, you will probably be most comfortable taking a number of actions simultaneously, in different directions. In other words, if you are experiencing a Uranian period—one in which you are restless and have an urge to be freer or to do things differently—you might examine various parts of your life and make small moves in several of them. For example, you might rearrange the furniture in your home, reorganize your routine at work, and in your social life go to a new place or take out a new person.

Such changes are not drastic, and if they do not work out well you can return to the way things were. Or, if any of them work out well, you can then make a further change in that area. Changes will occur suddenly but sporadically. Although it may seem that a great deal of action is taking place, meaningful progress may be slow. Eventually you may end up in a very different place from where you began, but you will do it gradually. This is because you might take one small situation and try a number of alternatives before you move on to the next small situation that needs to be redirected. In other words, you are always active, but sometimes you may move in circles instead of a straight line. Compare this to Uranus in Taurus, where movement can be so slow as to seem non-existent. Yet important changes will probably materialize at the same rate of speed that they will with Uranus in Gemini.

If you were to concentrate on only one segment of your life that seemed unsatisfactory and immediately make radical changes, you could feel restricted and anxious instead of freer. For instance, if, instead of rearranging your furniture, you packed up your belongings and moved to a new location with the first twinge of restlessness, you could regret what you had done. You might miss your old home and decide that your want to go back. Then you would spend a great of time and effort retracing your steps in order to return to your starting point. Whereas, if you take small steps forward without closing doors behind you, you might eventually move, but you would know by then that it was right for you. In other words, individuals with Uranus in Gemini are always ready for action, but the focus is more on movement than on sweeping change.

URANUS IN CANCER
(1948-1956)

With Uranus in the sign of Cancer we are linking the freedom-oriented and revolutionary qualities associated with Uranus to the nurturing, security- and home-oriented sign of Cancer. This seems an unlikely combination and one might expect discomfort with this placement. But there are always ways in which factors can be positively combined. And certain issues that were important in the US and the world from 1948 until the mid-fifties illustrate this. They also explain how people with this placement are likely to behave when Uranian matters are prominent in their lives.

Harry Truman became President in 1945. World War II was in progress and international matters took precedence over domestic issues. Shortly after Truman took office the war ended and he began to develop his own policy. His policy was called "The Fair Deal" and focused on **domestic reform**. It included such matters as the minimum wage, low income housing and Social Security benefits, all of which are clearly connected to security and taking care of one's own.

The war of the Uranus in Cancer period was the **Korean War** which **had parental overtones**. The United Nations forces were sent to protect South Korea (which represented the helpless child) against North Korea (the aggressor). In World War II, countries were defending their own territories, evoking the idea of self-preservation. In the Korean War the United Nations forces were protecting land which was not theirs—similar to a father or mother fighting his or her child's battles.

Another Uranian phenomenon of this period was the crusade launched by Senator Joseph McCarthy against what he viewed as **domestic corruption in the form of Communism**. In the early

1950s, as chairman of a subcommittee in Congress, he investigated more than 150 individuals from various walks of life to expose their controversial activities. They included people in government service, educational institutions, and the field of entertainment. Hysteria, fueled by the desire to protect the US on the home front, resulted in unfounded character assassination and a great deal of damage done to the lives of many people.

In 1954, he charged that Communists had infiltrated the Army. The Army-McCarthy hearings were televised and the tide of public opinion turned against Senator McCarthy. He fell from grace, was censured by the Senate within the next year, and left the public scene. All of this occurred as Uranus was in the last degrees of Cancer, and moving into Leo.

One more example that came to the fore during this period, and is appropriate for Uranus in Cancer, involved native Americans. In 1950, Dillon S. Myer was made commissioner of the Bureau of Indian Affairs. He began removing federal assistance from the tribes and started **selling Indian land**. Forty percent of the Indians became city dwellers and preservation of the Indian culture suffered as a result. Although there was strong objection to what he was doing, his policies remained in effect for the rest of the decade. And only after the Democrats came into power in 1960 were they changed.

This example shows that one's **roots or ancestry**, which are associated with the sign of Cancer, can be focal points in Uranian issues. It therefore provides some insight into factors that might be considered when someone with Uranus in Cancer is dealing with issues of freedom, individuality and revolutionary change. The impact of both Myer and McCarthy on society during the fifties illustrates another point as well. Emotions formed the basis of action and public response catapulted these men into prominence. Whether this prominence was fame or infamy depends on your personal viewpoint.

Because Cancer is a water sign, it makes sense that **emotions trigger Uranian activities**. Joseph McCarthy was able to rise to power by capitalizing on the public's fear of Communism. He tapped into the emotional climate in the country and evoked and fed on hysteria. **People with Uranus in Cancer** have the potential to react in the same manner as the general populace did when they were born. However, hysteria is not inevitable. With Uranus in Cancer, it is only a certainty that matters involving freedom, individuality and revolutionary changes trigger your emotions. And you will probably use your feelings and instincts in resolving such issues. You need not, however, be frantic or out of control in your response.

Even if you are an extremely down-to-earth person according to your horoscope and your behavior, with Uranus in Cancer, you will most likely not need specific examples in forming decisions connected with Uranian matters, nor will you be likely to respond to the common sense approach. Your feelings will guide you, and it is through appeal to your emotions that others can reach you.

If you understand and accept what needs to be considered when you are dealing with Uranian issues, you might be able to respond faster and thereby avoid the impatience and restlessness that can be connected with Uranus. The self-centered quality of Uranus will be modified by the desire to nurture and protect others. Don't waste your time wondering why you are not thinking of yourself first. It is not enough for you to say "I want it" or "I need it." It is easiest to express your individuality if you are convinced that it is in the best interests of other people. And you can use unusual means to move in new directions if the welfare and security of someone else is involved.

You will probably align yourself with unusual people who you consider to need nurturing. They may look or act in a manner that does not bring out sympathy in other people, but they touch your mothering instincts. So you begin to take care of them. But you do not want to help them forever. If they hang around for too long, you become impatient. You may begin to feel trapped and want to rid yourself of them. And it is possible that your feelings keep building until one day you suddenly end the relationship with a display of emotions. You can probably avoid this occurring if you set a time frame (preferably a rather short one) for tending to these individuals. If you have not nurtured enough when the deadline arrives, you can always extend it. But, you will feel less trapped if you know the end is in sight.

Since Uranus is also a planet of creativity, your emotions need to be aroused for you to be inventive or to execute something artistic. You will also use your intuition to guide you. The results of your efforts could come in the form of ideas, but you might also produce something quite tangible and practical. It is only the procedure that is based on feelings.

URANUS IN LEO
(1955-1962)

An appropriate key word for **Uranus in Leo is dramatic**. Chances are that even the most covert actions will not remain hidden for long. Evidence for this in the mundane world can be seen by the way in which John F. Kennedy, who was President of the United States during part of this period, dealt with Uranian issues during his term of office. Another example of this is Martin Luther King, Jr. and his part in the Civil Rights Movement.

Kennedy was considered a "Media President." He used television, radio and the press extensively during his presidency. Two excellent illustrations of his use of the media were his handling of the Bay of Pigs incident and the Russian missile crisis in Cuba. In both cases he came before the public through television. He shouldered blame for the failed invasion in the first example and publicly threatened the Russians in the second.

King, too, was not shy when it came to publicity. His nonviolent approach to the Civil Rights Movement had a great deal of coverage in all facets of the media. In fact, the media were an integral part of the movement and played an important role in achieving some of the improvements in race relations that we see today. There is still more to be accomplished, but without the exposure of racial injustice through the mass media, blacks might still be sitting in the back of buses in the south, and denied entrance to some restaurants and other places of business in many parts of the country.

Besides making use of publicity and seeming to enjoy the limelight, these two men shared another quality that we can associate with Uranus in Leo—idealism. Kennedy had an optimism that was contagious. Even the dangerous step he took in the Cuban missile crisis was not condemned by the public. Although there were those who were opposed to his taking this step, the attitude held by a large segment of the general population was that it was the right thing to do, and would work. King's idealism was evident in the cause with which he aligned himself and in many of his speeches. Probably the best remembered of these is the "I had a dream" speech.

Both men were talented performers and used their abilities to gain the support necessary to achieve their goals. Although each had some success in their causes, neither lived to see their dreams come to total fruition. It was as though they set the scene while Uranus was in Leo. They dramatically presented the issues that were important to them and the reforms they wanted to effect. But it was not until Uranus moved into Virgo that serious, tangible results were evident. As in the beginning of the century, injustice was exposed when Uranus was in a fire sign, but it was not until the planet moved into an earth sign that meaningful change took place.

Kennedy and King both died when Uranus was in Virgo. And it may be that their deaths (which were unusual and unexpected, and therefore, could be considered Uranian) partly led to the ultimate accomplishment of their goals. Perhaps their dramatic exits were the harkings back to Uranus in Leo, even though they took place after the planet had moved into Virgo. Kennedy's demise, added to Lyndon Johnson's ability to practically negotiate, may have helped Johnson to get some of Kennedy's "New Frontier" programs through Congress. (Even the title "New Frontier" sounds like Uranus in Leo.) And King's assassination, which made him a martyr for his cause, is still a stimulus for the Civil Rights Movement.

If you have Uranus in Leo, you probably are not subtle in regard to revolutionary issues or matters of individuality and freedom. Even if your Uranus is in the twelfth house, your revolutionary tendencies will frequently erupt externally. You might try to be covert, but, as with the failed invasion at the Bay of Pigs, your actions usually somehow become public knowledge. This should not disturb you too much because you want to let the world know of your intentions. It is a way to get support and recognition from others. So you might not be the best choice for highly secretive, undercover work. You can, however, achieve your goal with your open approach. You can stand on stage and dramatize your cause so that others are stirred into action.

The reasoning behind your revolutionary actions and expression of individuality is based on idealism. You will probably not rebel unless you believe your cause is just. And once you have taken your stand, you may be surprised if everyone does not agree with you. Disagreement may occur because you have not worked out the practical details to accomplish your goals. It is the end result that is most important, and it may not always be easy to convert your dreams into reality.

Since you expect everyone to agree with your views, you initially believe that your ideas will gain momentum and succeed without effort. However, when you meet with resistance, you are not only surprised, but you could feel thwarted as well. Having contingency plans, just in case, may be helpful. But there are other choices as well. You could keep talking about your plans, in hopes that you can still rally needed support which is a definite possibility with your ability to be a motivator. Or you might bottle up your feelings and suffer (an unlikely possibility when the sign of Leo is involved). Or, you may ally yourself with people who have Uranus in Virgo (or have earth signs strongly emphasized in other ways in their horoscopes) and, thus, can help to make your dreams materialize!

The modern technology that was in the foreground during the period when Uranus was in Leo can give insights into the inventiveness and creativity of individuals with this placement. **Nuclear energy** was highly emphasized. Two atomic energy plants to supply electrical energy were approved to be built. Nuclear submarines were being used, and underground nuclear testing was begun in Nevada. This was also a time when the Advanced Research Projects Agency was set up by the Defense Department to promote space exploration.

There is nothing subtle about nuclear energy. Even underground explosions are not easy to cover up. Therefore, creative projects that individuals with Uranus in Leo undertake need to be dramatic, or at least noticeable. There is little excitement about being involved with creativity that is behind the scenes and creative juices will flow as long as the enthusiasm is maintained.

The promotion of **space exploration** adds the possibility of another quality that is necessary in Leonine creativity—expansion. This combines challenge with the desire for drama and explains why the "nuts and bolts" type of creativity would be avoided. It is interesting to note, however, that humans did not step onto the Moon until Uranus was in Virgo.

CHAPTER 6

URANUS IN SIGNS: VIRGO, LIBRA, SCORPIO

URANUS IN VIRGO (1961-1969)

As with all the earth signs, you would expect the revolutionary and creative activities of Uranus in Virgo to deal with **practical issues and involve tangible, concrete results**. With the sign of Virgo specifically, there should be attention to detail, and the pace of progress would be slow in spite of the sudden and erratic qualities usually associated with the planet Uranus. Yet, when results materialize, changes may seem sweeping and abrupt. Once action is taken, all of the necessary ground work and analyzing will probably fade into the background.

For example, in the month following Kennedy's assassination, **a large number of laws were passed to wage war on poverty**. Kennedy was in office from January, 1961 until November of 1963. Early in his administration he declared a "war on poverty." He had numerous ideas in regard to changing US policies, but it was only after his death that many of these ideas were enacted into law. This may be attributed partially to the martyrdom of Kennedy, but also the "wheeler-dealer" capabilities of Lyndon Johnson.

With Uranus in Virgo, preparation to make changes may have taken time, but the ultimate results were sweeping and only seemed sudden.

The revolutionary changes in the laws that were passed in the month following Kennedy's assassination are the type to be expected with Uranus in an earth sign. They included **protecting the voting and housing rights of Blacks, the establishment of medical care for the elderly, and programs for the financially disadvantaged**. There was also emphasis on improving education. All of these issues are down-to-earth, and health matters are particularly appropriate for the sign of Virgo.

Even with the legislation passed, however, more work had to be done. The battle for Black civil rights is an excellent example of progress being slow in earth signs (and slowest of all in Virgo). The Civil Rights Act was passed in 1964. Schools were supposed to be desegregated. Housing and employment opportunities were supposed to be open and equal. But the new laws were not always followed and protests began.

In 1963, 200,000 people marched in Detroit to protest discrimination. After that, half the Black children in Chicago boycotted the schools to point up the fact that desegregation was not being implemented. Then the protests began to take on a more violent tone. In 1964 there was a riot in Harlem, in 1965 in Los Angeles and in 1967 a series of race riots erupted across the country. After Martin Luther King's assassination in 1968, another series of riots ensued.

Even though Uranus is the planet of revolution, you would not expect a great deal of violence when that planet is in Virgo. This may be explained by the fact that **Pluto was conjoining Uranus in Virgo from 1963-1968**. (This will be discussed further in the chapter on aspects.) Comparing Uranian events from 1961 to 1963 and in 1969 with those that occurred during the years of the Pluto-Uranus conjunction (1963-1968), might suggest possible differences between these two groups of people born during those respective years.

First and foremost, all the riots took place while the Pluto-Uranus conjunction was in range, supporting the idea that there was more volatility during that time than the rest of the period in which Uranus was in Virgo. But even with the conjunction, violence was not the immediate response. Although the riots seemed to erupt suddenly, they occurred only after peaceful protests were tried and failed to have an impact.

Anyone with Uranus in Virgo wants to take the most practical and efficient path in order to make tangible changes. They will usually work out the step-by-step plans before they launch their first attack. And, as evidenced by the timing of the riots, even with the Pluto-Uranus conjunction, action was not taken quickly.

The desire to investigate and get to the bottom of a situation which we connect with Pluto, combined easily with the slow-moving thoroughness of the sign of Virgo. The potential volatility of Pluto was temporarily delayed by the need of the earth signs to take the practical approach—the one that will work. And Virgo also modified the erratic and impatient qualities we associate with Uranus. At first glance, you might expect individuals with Pluto conjunct Uranus always to be ready for a good revolution in which the old regime would be ousted simply for the sake of transformation. But this really is not the case. With these planets being in Virgo, there need to be sound, practical reasons for a revolution. So those with the conjunction, as well as those with Uranus in Virgo but not conjunct Pluto, would tend to bide their time and investigate the various alternatives for change available to them. Both groups would probably try the most conventional or safest paths first. However, if either group finally resorts to a revolution, those with the conjunction are likely to take the more drastic actions and make the more sweeping changes.

The **Vietnam War**, the major war of the Uranus in Virgo era, also exemplifies characteristics connected with the sign of Virgo. It was, without doubt, the most unpopular war in U.S. history. The United States' involvement in Vietnam began during Truman's presidency, but it was not until 1964 (during the Pluto-Uranus conjunction) after a North Vietnamese attack on two American ships that large numbers of American troops were sent there. The war escalated over the next few years and it became less and less popular.

What was particularly offensive to the American public were stories about the killing of civilians in routine bombings and tales of corruption and drug use among the American military. It was considered an **immoral war**, and morality is an important key word for the sign of Virgo. Applied to the individual with Uranus in Virgo this adds the need to examine the morality of one's stand before revolutionary action can be taken. This further explains why individuals with this sign placement of Uranus (even those who have Uranus conjunct Pluto) do not take action more quickly.

Let us now look at the **modern technology** of the Uranus in Virgo period to get information about the creativity of those born during this time. One advancement that seems to epitomize Uranus in Virgo stands out. This is the development of the **microchip**.

Although the microchip made its first appearance in 1958 while Uranus was in the idea sign of Leo, it was not until 1962, when Uranus was firmly entrenched in Virgo, that it was mass produced. But, aside from making this valuable discovery available to more and more people, there was another quality that reflected Virgo. We know that Virgo pays a great deal of attention to minute details and **throughout the 1960s the microchip became more efficient and more compact**. "In 1964, for example, a chip a tenth of an inch square contained a total of 10 transistors and other components. By 1970, no fewer than 1,000 components were crammed into the same-sized chip, at approximately the same cost as before."[2] In this quote, both the painstaking patience and attention to detail are easily ascribed to Virgo.

Therefore, in terms of creativity, **individuals with Uranus in Virgo should look for projects that may take patience, but can also produce useful results**. Cost may be a consideration, but this factor needs to weighed against the worth of the product. The more practical and useful its application, the less likely expenditures will be a concern.

---

2. *Computer Basics*, Alexandria, VA: TimeLife Books.

URANUS IN LIBRA
(1968-1975)

Since Libra is the sign of **peace and fairness** it would be expected that Uranus in that sign would be concerned with these matters. If any kind of revolution were to occur, injustice would be the grounds. During the Uranus in Libra period, the **war was ended in Vietnam**. There was no winner of this war, but then, that wouldn't be the point with this placement. The desire of Libra is peace, not necessarily victory. The Uranian change in this case was the cessation of violence.

This does not mean that there was no violence during this period. There were the incidents at **Kent State and Mississippi State in which people were killed**. But usually in occurrences such as these, **peace and justice were the bases for the uprisings**. If the cause is justified, any means are allowable. Even a worthwhile goal can trigger brutality.

Protests, however, need not always be violent nor generate violence. There was a turning point in August of 1969 (with Uranus just entering Libra) when protesters became more nonconfrontational. The event that took place at this time was a three-day rock concert called "Woodstock." More than 300,000 young people attended. Food was in short supply and sanitary conditions were inadequate, but the emphasis at Woodstock was on peace and love. Violence was expected, but did not occur. **Woodstock became the symbol of the counterculture movement**.

It is interesting, too, that Libra is the sign associated with **the arts** and it was music that characterized Woodstock the most. This does not necessarily mean that individuals with Uranus in Libra must use the arts in connection with revolutionary action, but it is possible. And if you have this placement you might want to consider incorporating the arts in some way when you are making changes.

On the political scene during this period, the **emphasis was on peace and diplomacy**. The cessation of the Vietnam War exemplified the focus on peace. In regard to diplomacy, President

Nixon took a 12-day trip around the world in August of 1969 (with Uranus in the early degrees of Libra) and let it be known that he was willing to negotiate with Communist nations. While in Romania during this trip, he said "nations can have widely different internal orders and live in peace." In 1972 he paid an historic visit to Communist China. He described it as a "journey for peace." This trip led to a cultural exchange and diplomatic relations between the two countries.

During the Uranus in Libra period, another event occurred which is far different from the overt diplomatic overtures for peace. This event was the **Watergate break-in**. Five men were arrested in an attempt to steal political material at Democratic National Headquarters. The purpose was to have influence over the Democratic Party's selection of candidates. The break-in ultimately led to Nixon's resignation, when it was proven that he was aware of this covert action. The Watergate incident points up another Libran tactic that is often overlooked—**covert strategy**.

The desire to keep the peace could be stated as the reason for avoiding direct confrontation. Libra is a sign that **prefers not to be involved in unpleasantness**. The Uranus in Libra person may have the revolutionary ideas, but rebellious activities are usually left to others. Therefore, with Uranus in Libra, the key to successfully making dramatic changes or starting a revolution, is to subtly convince someone else to carry out your plans. This is done partly to maintain the image of the peacemaker, but also because it is easier to create ideas than it is to take physical action.

If you have Uranus in Libra, a comfortable way to proceed should be obvious from the preceding paragraphs. **When contemplating change or expressing your individuality, you would first want to think about it**. Next you might consider an amiable way to get things started. If this fails, then you should figure out how to motivate someone else to take the initiative to reach your goal or help you express your individuality.

You will not have to look far if you have a partner with whom you regularly share your revolutionary thoughts as well as someone to take action for you. **With Uranus in Libra you need not stand alone fighting for a cause**. It is much more satisfying to have company on such a project. You also need not be the leader of the group because equality is better than being either the superior or the subordinate. Even in your search for personal freedom, it is probably more gratifying to share the quest with someone else.

In terms of creativity, it was already mentioned that the arts might be used in connection with revolutionary action or protests against the establishment, as was illustrated with the Woodstock rock concert. It is one way in which you can voice objections to the establishment and express your revolutionary ideas without violence.

The placement of Uranus in Libra could also mean that you have a **unique, nonconforming approach to the arts or perhaps you use modern technology to express your artistic talents**. If you are artistically inclined, what you produce may be considered ahead of its time, or at least not in step with present standards. In this age of computers I am not surprised that a number of my clients who were born with Uranus in Libra are drawn to the graphic arts to express their creativity. At this writing they are still relatively young. Therefore, it is too soon to know what impact they will have in the world. But if we look to the preceding period of Uranus transiting Libra (1884-1890), one excellent example of **someone who both displayed originality and used modern technology was Charlie Chaplin**.

URANUS IN SCORPIO (1974-1981)

As Uranus moved into Scorpio the focus was no longer on making peace. The last of the US troops were evacuated from Vietnam at the end of April 1975 but this was just as retrograde Uranus was about to reenter Libra for the last time. Then other issues came into the foreground. The most prominent of these involved **covert activities**.

One such matter was concern about the role the CIA might have played in the **death of former Chilean President Salvador Allende Gossens** and the overthrow of his government. The Senate Foreign Relations Committee ordered an investigation into these allegations. Another incident was the seizure of a US cargo vessel by the Cambodian government with accusations that the ship was part of a **United States spying operation against Cambodia**. Still other examples were the establishment of a Commission to examine charges of domestic spying by the CIA, and the Senate Select Committee on Intelligence Activities looking into heretofore concealed questionable activities by the FBI, IRS, and the US Army, as well as the CIA. All of these investigations dealt with spying and ferreting out hidden information—appropriate concepts for the sign of Scorpio.

Two other newsworthy items that occurred during that period, and were appropriate for the meaning of Uranus in Scorpio, were the Iran Hostage situation and the case of Patricia Hearst. On November 4, 1979 a group of Iranian students stormed the US embassy in Teheran and captured 66 Americans. The students were protesting the Shah of Iran having been allowed to enter the United States for an operation. They wanted the Shah returned to Iran to face trial. In retaliation for the hostage situation, US President Jimmy Carter deported Iranian students who were illegally in the United States and froze Iranian assets that were in American banks.

Although these measures failed to produce the return of the hostages, both the action of the students and the response of the President illustrate ways in which people might act with Uranus in Scorpio. The captors did not openly remove the Shah from the hospital. Instead they took the circuitous route and tried to manipulate the Americans into returning the Shah. The response of the President was one of "I'll show you." The **freezing of other people's money** also seemed appropriate for the "getting even" attitude of Scorpio.

The other excellent example of Uranus in Scorpio mentioned above was the story of Patty Hearst. The events that occurred in her life during this time speak for themselves. Without elaboration they evoke keywords for Uranus and Scorpio. She, like the hostages in Iran, was **abducted**. Her kidnappers called themselves the Symbianese **Liberation** Movement. She was allegedly **raped**, kept in a **dark closet** and eventually was **brainwashed** into joining the group. She was **transformed**. She took on a new identity and even looked different than she had before. If someone were asked to create a work of fiction that would epitomize Uranus in Scorpio, they could probably not do better.

This does not mean that everyone with this placement would behave as Patty Hearst's captors did. This example is rather meant to point out that there are two general routes the revolutionary with Uranus is Scorpio would likely follow. The first would be to convince, and/or perhaps manipulate others to join in their revolution. This route would also include the covert behavior such as the type alluded to, not only in the Patty Hearst story, but in several of the other examples. The second route is the more overt one associated with the volcanic eruption quality of Scorpio. You take visible action on a large scale. This is illustrated by Patty Hearst's alleged involvement in bank robberies with her kidnappers, the assassination of the Chilean President and overthrow of his government, and the Iranian seizure of the American embassy in Teheran.

These two routes need not be exclusive of each other. In fact, although they may seem diametrically opposed, the second path may follow the first, and is probably the most comfortable procedure for a Uranus in Scorpio person to follow in order to make changes and/or express individuality. **If you have Uranus in Scorpio, you might initially investigate the changes you want to make quietly, without fanfare**. You can elicit support in any manner you find appropriate as long as you do not publicly announce it. Once you know that others are on your side you will feel secure enough to take drastic, definitive action.

We have now followed Uranus in its movement in the twentieth century and discussed it in all the signs. The historic events given above are just a small number of those that might be discussed. They were described to illustrate the trends of the different periods. If you delve further into what was occurring in the world at your time of birth, you might uncover more information that will be applicable to your own pattern of behavior and possibly derive some more ideas as to how build upon it or improve it.

CHAPTER 7

URANUS IN HOUSES 1, 2, 3

The transpersonal side of Uranus has been shown through discussion of mundane events connected to the planet's movement through the signs. This information became more personal as it was applied to the individual. However, the **planetary messages become even more personalized when we consider the house placement of the planets**. The house in which each planet is posited is where the themes connected with it can be expressed most directly.

Since Uranus is the planet that represents our revolutionary tendencies, eccentricities, independence and creativity, especially mental creativity, the house in which it is placed is where these motifs should be most evident. The individual needs to **manifest creativity and individuality, wants to be free, and may be nonconforming or break with tradition in that area**. Or that person could start a revolution if Uranian requirements are not otherwise met. Even if you never start a revolution you will probably experience **restlessness** in the house in which Uranus is placed, and you may institute or have to deal with **abrupt change** there.

There is, however, another quality connected with Uranus that must be dealt with and incorporated in order to make good use of Uranian themes. This is the need to define boundaries. How can you be different in an area if you do not know what the norm is? Or how can you determine what you want to change if you do not grasp the factors that you are dealing with? Astronomically, the fact that Uranus rotates on its side may illustrate its eccentricity. The rings that were discovered around the planet, however, indicate the need to set boundaries (as with Saturn). Therefore, **you may be unconventional in the house in which Uranus is placed**, but you have to become familiar with the limitations that are there as well.

URANUS IN THE FIRST HOUSE

If you have Uranus in the first house, Uranian qualities will be strongly exhibited in your personality. So when people describe you they will include such words as **impatient**, **original**, possibly **eccentric**. There is also an **objectivity or aloofness** that will likely be prominent in your character. This may attract those who want an impartial opinion because you can understand everyone's point of view. You will usually be fair and not take sides.

This detachment can be a self-protective device as well. If someone interferes for too long with your need for freedom and becomes too demanding of your time or space, inwardly a veil will come down separating you from the other person. Outwardly it will be perceived as an impenetrable wall. Once this wall is in place, your attitude toward that individual will be cold. And the more your emotions are appealed to, the more detached you can become.

Another facet of the **detachment** is that you do tend not to be judgmental. You have a *laissez faire* attitude. If another person does not try to interfere with who you are or what you believe in, you will reciprocate. You will accept that individual as is and not try to place your standards or values on him or her. Most of us would mention an unusual characteristic about someone but the Uranus in the first house person might not even notice it, let alone mention it. For example, if someone without Uranus in the first house were describing an individual who had green hair, he or she would tell you that this person had green hair immediately, because it is not the norm in our society. The Uranus in the first house person would not think of using such standards. There might be an awareness of green hair being different, but it would be accepted as an integral part of that person being described.

On the other hand, **if other people try to inflict their standards on you** (which you would never do to them), **you would probably develop a distaste for their company and most likely try to avoid them as much as possible**. I have a client with this placement and I have seen this reaction a number of times in the years I have known her. She seems to like and accept everyone who does not interfere with her, but has no time for people who try to control her. However, I have also noticed that if the interfering people become more accepting, my client usually mellows a bit toward them. This shows that those with Uranus in the first house can change their attitudes. It should be added that the change may take time, but at least it can occur.

The **impatience** you experience with Uranus in the first house may be mental or physical. You could have difficulty sitting still or you might have a short attention span. So although you can be cool, objective, aloof, you will not be seen as sedentary. An air of electricity and a flurry of activity frequently surround you. Even if you are not constantly on the move, your active mind is evident. You might sometimes say that you wish life were more peaceful, but you can become easily bored. In fact, if your environment is too serene, you will probably stir up some excitement. Therefore, the flurry of activity that is always in your vicinity may be self-generated.

Since the first house represents physical appearance as well as personality, your **nonconformity may be apparent in your dress or in the way you wear your hair**. So you may be individualistic in your behavior, and/or the way you look. You could even take pride in your nonconformity and independence whether it is in your actions or in your dress. If you think that you are not expressing the independent side of your nature, you might consciously try to change your appearance before you take more drastic action. If you make great changes spontaneously, without thinking about the consequences, you could end up feeling disoriented rather than freer.

Accepting the facts that **you enjoy being different**, that **you need excitement** and that **you want to be creative**, is essential to make the most of your first house Uranus. You don't have to waste your time looking for excuses for your uniqueness.

In order to capitalize on your individuality, you have to determine what standard behavior is, and/or what the fashionable trends in dress are. As you examine what is considered acceptable behavior, you can determine what you want to conform to and what you would like to change. You need not do everything differently from others. You may need your freedom of expression, but if you are too bizarre, people may not take you seriously. If you really want to make a difference, you could have a greater degree of credibility and a stronger impact on changing the standards of acceptable behavior by conforming to the parts of present standards that are agreeable to you. This is true of dress as well. If you are aware of the styles most people are wearing, you can make noticeable changes but not be so different that you will be considered weird.

In regard to the **possible short attention span** connected with Uranus in the first house — you need not consider this a flaw in your character. Do not try to force yourself to spend long hours concentrating. It will probably be frustrating if you do. Instead, capitalize on it. Make sure you have a variety of activities that you can alternate with each other. You will accomplish much more by doing this than worrying why you cannot spend many hours on a single task.

Your **need for personal freedom** is another fact that you have to recognize. Recently a client with this placement came in for a reading and asked why she was unable to maintain close friendships for long periods of time. It was an issue that concerned her and she wanted to know what was wrong with her. There were other factors in her horoscope that indicated she needed closeness as well, otherwise, she might have been quite content being a "loner." The pattern she had established was that she would meet someone she liked, found herself spending more and more time with that person and then began to feel restricted. So, somehow the two people would have an argument, and my client would be free again. This cycle had prevailed for a number of years.

When she first became aware of the pattern she blamed the other person, but when the scenario was repeated several times she began to know that she must be contributing to the problem as well. The solution was not really that difficult. Once she understood the dilemma, she accepted the idea that she needed both togetherness and time alone. And now she arranges her schedule in such a way that if she spends a long period of time with someone on one day, the next day she will seek solitude.

You should avoid such things as placing yourself under the domination of another individual, or taking on long-term commitments that make you feel trapped. You should take pride in your individuality. If you seek ways to express your creativity and find areas in which you can have space and feel independent, you can make the most of your first house Uranus.

URANUS IN THE SECOND HOUSE

There are a number of ways in which Uranus in the second house can manifest. Since this area may represent the way in which you earn a living, **you will probably need a great deal of freedom in your work**. I have a number of clients with this placement who are in business for themselves. This certainly helps to avoid feeling suppressed by tyrannical employers. It does not, however, automatically provide the native with a sense of material security.

Money may come in spurts, and could flow out in the same manner. People with this placement are not usually patient shoppers. They can spend spontaneously on a whim and later repent. What I usually suggest, since income can be sporadic, is that when a windfall occurs, some money should be put away and considered gone when shopping. An IRA, a Certificate of Deposit, or some other form of **enforced savings in which funds are not easily retrieved should be instituted**. Of course, money can be removed from such accounts, but this requires effort and Uranus suggests that there isn't the patience to take care of all the details. So it is likely that the funds will remain untouched.

Time and again clients who have followed this advice tell me that the money they save gives them a sense of security, and makes them feel more independent. One woman with Uranus in the second house, who is a graphic artist, started her own business about three years ago. She was a little nervous about being on her own, as is common with anyone starting a new business. As she presented her ideas to perspective clients, she was concerned with what she considered her lack of originality. She thought of herself as being creative in the positions she had previously held in established companies and began to worry that perhaps she wasn't talented. She felt that the jobs she did get were because her bids were low. However, once she began to earn money and to put some of it away, her creativity magically reappeared. She began to present less conventional ideas and business has picked up. This example reinforces the premise that when you feel secure, you can be freer.

You need not, however, be self-employed with Uranus in the second house. You can work for someone else if that individual does not try to make you conform to inflexible standards, nor attempt to direct your creativity. If you are allowed to **express your individuality in your occupation** or are encouraged to use your originality, you may have no complaints about your job at all. You may still be erratic about spending money and, perhaps, not feel pressured to save. But if you are in someone else's employ you need not worry about where the next paycheck will come from.

Other possibilities with this placement that you might already use, or you could consider as substitutes for less desirable alternatives, are to have unusual hours on your job or change your schedule frequently. Or you might work on commission, or choose to work part-time, or possibly have more than one job so that you will not get bored. Instead of being erratic or too spontaneous in your spending, you might purchase Uranian items such as computers that could be a worthwhile investment rather than a frivolous expenditure. And finally, you might select a Uranian occupation such as electrician, computer technology, or become involved with a humanitarian field or cause.

You could think that it would be nice to be independently wealthy and it would be reassuring to have financial security. You might even say that it would be wonderful to be able to be free of money worries. And, after all, freedom is a key word for Uranus — security is not. Should you feel too secure or force yourself to be frugal (which is highly unlikely), you might find that something is missing. It could be that you miss the challenge of making ends meet or the exhilaration you feel when you take a risk and spend money spontaneously.

Since this house also represents your own feelings of **self-worth**, Uranus placed there could mean that your opinion of yourself could fluctuate (as with the graphic artist mentioned above). When you are not feeling good about your identity, you might sabotage yourself. You might exhibit bizarre behavior that could get you fired, or you might go on a spending spree that could put you deeply in debt. Then you will feel worse about yourself than you did before. Instead, define your situation. Try to determine why you are feeling limited or insecure. See if you can make small changes that could lead in a new direction and eliminate obstacles to your independence and creativity.

URANUS IN THE THIRD HOUSE

Uranus in the third house can indicate that you have **unusual sisters and brothers**, and/or that you have fluctuating relationships with your siblings. There isn't much you can do about it if they are different from the norm, except maybe to accept it and realize that this placement can also mean they are creative as well. Ted Kennedy, who has Uranus in the third house, certainly had siblings that fit both descriptions.

If you have Uranus in this house **do not expect constant, intimate connections with your brothers and sisters**, especially the oldest sibling. Too much togetherness will not work. You should also look elsewhere if you need continuous help or compassion and accept the realization that you don't want to give your sisters and brothers constant help and attention either. You will be objective about them and if you keep some distance between you, at least periodically, you need not have problems with them. If you acknowledge all of this, you can avoid disappointments with your siblings' performance, or their interference with your freedom. If you associate with them in order to share creativity and originality, and don't anticipate a great deal in return, your interaction can be very gratifying.

Since the third house is also the house of **communication**, Uranus in the third house can mean that you are quick to speak. So you may be considered outspoken, even to the point of being tactless. Your objective Uranus will be more focused on freedom of speech than the impact upon the person with whom you are talking. You may think about the fairness of what you are saying, but not consider the sensitivity and feelings of those on the receiving end.

Ideas may come quickly and evaporate just as fast. Therefore, you might choose to type them or put them into a computer, or speak them into a tape recorder. When it comes to listening, you might not always have the patience to allow someone else to talk for too long, especially if that individual speaks slowly. You frequently may anticipate what the other person is trying to say and finish sentences if that individual is not getting to the point quickly enough.

Even though you have your own ideas and do not hesitate to express them, **you respect the right of others to their opinions**. You may actively debate issues but will not force your views on others. If, however, others try to make you conform to their opinions, you will not only resist, but might even become more outrageous with what you say. This does not necessarily mean that you believe every opinion you voice. Sometimes you simply enjoy playing Devil's Advocate. It is one way to determine what is right. You don't like to be told how or what to say or think. However, if your opponent offers a convincing argument for a particular point of view, you could change your mind. You may stick to your guns until the end of the conversation, but in the next discussion you have on the same subject with someone else, you may espouse your former opponent's thoughts as though you had originated them. So others may see you an inconsistent, but you know that you are really a free thinker.

Since this area also represents your **immediate surroundings**, you might choose to live in a Bohemian neighborhood or find that your neighborhood itself is not unusual, but your neighbors are either strange or creative. With aloof Uranus in the third house, however, if they are strange you shouldn't have difficulty keeping your distance.

Finally, the third house also represents **elementary education**. If you are an adult reading this book, you might remember that you were a rebel in grade school. But more importantly, if you have a child who is in elementary school and has this placement, you might be a little more understanding of the child's behavior in school. Or better yet you can help him or her to find ways to express individuality or creativity. A progressive type of school might be better than a traditional one. Or it might be a good idea to look for teachers who are stimulating. Then the child will feel challenged and creative, and may not need to start a revolution. And you won't have to worry about receiving those scathing notes or calls from the principal.

CHAPTER 8

URANUS IN HOUSES 4, 5, 6

URANUS IN THE FOURTH HOUSE

Traditional astrological text books will tell you that Uranus in the fourth house could indicate that you **change your residence** frequently. So what is brought to mind is someone who travels from place to place establishing home after home. And this can happen. You can be a world traveler or do as one of my clients with this placement did. He moved every three or four years as he was growing up, but did not go very far. He simply moved across the street to his grandmother's house for a few years, then back to his parents' house, then back to grandmother's, etc. He has, however, made more distant moves in adulthood. But, it also is possible to have this placement and stay in the same house all of your life.

If you never move, however, chances are that you came from an **unusual home environment, had an eccentric or creative parent, or simply that your adult home will differ greatly from the home of your childhood**. Since the house in which you find Uranus in your natal chart is where you may consciously or unconsciously break with tradition, you will most likely operate differently from the way your family of origin did. If there were strict rules and regulations in your childhood home, you might carefully avoid this in your adult home, and have a rather unconventional household. If your childhood home was unusual or nonconforming, then you might crave stabilization and create a very conventional environment.

It is also possible that you would **move from one extreme to another**. If your childhood home was restrictive and you create an unconventional home, after a while you might miss the predictability or the childhood traditions. Then you might inform those with whom you live that this is the year you are preparing the Thanksgiving turkey and inviting the relatives to dinner to re-establish the old family custom. But after a year or two of this you could tire of it and revert to your previous stand of "who needs tradition?" In other words, there could be a number of changes in attitude in the home in the course of a lifetime. Prior to each change, however, there will probably be an analysis of the behavior in the home. You should consider what is restrictive to you and change that segment. Otherwise, those with whom you live might become confused.

One possibility that might be satisfying to you and not create confusion for your housemates, is that you can regulate certain parts of your household routine (giving yourself permission to change them when you want to) and remain spontaneous and free in other aspects of the home. Another possibility is that you establish a firm home-base and periodically remove yourself from it by **traveling**. In this way you are free of the home while you are away, and yet it is there to return to.

Since the fourth house also represents **one of your parents** it might be that one parent was erratic and at least partially responsible for the instability in your early home (if there was instability). Or your Uranian parent may have been creative instead of (or along with) being inconsistent, and contributed to an atmosphere in the home which fostered individuality. When I think of this placement, I am reminded of the play, *You Can't Take It With You,* in which those who lived in the house were involved in their own activities and unusual, interesting people wandered in and out.

Even if you live in what you consider a totally **free environment**, you may not be quite as spontaneous as you think. Although you can talk about your flexibility in the home and really believe it, you probably do not react well to surprise visits. You might initially react in a gracious manner and pride yourself on your ability to respond so spontaneously. But after a short period of time you might begin to resent these unannounced visitors because they interfere with your freedom. You might feel that you have to entertain them and are thus restricted. Of course, you could display your aloofness and ignore them. But even if this occurs and you do not let them keep you from doing what you want to do, their presence serves no purpose.

You probably prefer to determine how and when you are going to be spontaneously sociable. You might not only tell your guests when to arrive, but also what time to go home. In fact, if you decide to have a party, you will probably feel most comfortable if you specify the time of arrival and the time of departure. This is not essential, however, because if your guests do not leave when you want them to, you could become so cool and detached that they will probably get the message to go home rather quickly.

Your **unconventionality in the home** may take a totally different direction. You might have unusual taste in furnishings or color schemes. Your decor could range from the bizarre to interesting and eye-catching. Your furniture might be constructed of glass and steel or in some other way be futuristic in style. Or you could combine different periods of furniture. But whatever you do in decorating will be distinctive and noticeable.

It may seem as you read this section on Uranus in the fourth house, that there are conflicting messages. You could yearn for a stable home and work very hard to establish a secure base. Then you might complain about being bored and/or do all you can to disrupt it. Another seeming conflict could be in regard to the parent described by this house. You could have considered this parent to be original, creative and independent, or erratic and unreliable. It is impossible to say from the placement alone whether you admired or disliked this parent, but you probably will have seen more than one of the qualities just mentioned.

If you have this placement and experience the inconsistency, the first step is to realize that there is nothing wrong with you. Tell yourself and anyone who complains about you that you are inconsistent and that sometimes you need structure and sometimes you need freedom. Although you want to express your individuality, making drastic changes too quickly could disorient you. Analyze your home situation. Determine what you are dissatisfied with and make small changes first. This could satisfy your urge for change or lead to bigger and better changes.

URANUS IN THE FIFTH HOUSE

With Uranus in the fifth house, you may consciously or unconsciously try to **raise your children differently from the way in which you were raised**. Or you could have original, creative and/ or rebellious children. Another possibility is that you might be inconsistent in the way in which you handle your children. You will undoubtedly say that your want your children to be independent, but probably not totally nonconforming. So if they become too defiant or too different, you could clamp down on them. This again demonstrates the need for defining or setting limits on the freedom of Uranus. I heard one parent with this placement say to her child, "Of course I want you to be your own person. I just don't want you to be a total misfit."

You may also have an **objectivity in regard to your children**. So you are able to look at them in a detached manner and it may be difficult for you to be emotionally demonstrative with them. This does not mean that you are uncaring, just that it could be hard to express your feelings.

The fifth house describes the way in which you deal with children in general, and specifically your own (if you have any). And this may be especially true of your first child, but possibly somewhat softened for your subsequent children. Your attitude toward your second child may be modified by examining the seventh house; your third by the ninth; your fourth by the eleventh, etc.

Another definition of the fifth house is the **house of creative self-expression**. Producing children is, of course, a form of creative self-expression, but this area also describes the way in which you operate creatively and what some of your talents may be. With Uranus in the fifth, your creativity probably comes in spurts; so there could be times when you can't record your ideas quickly enough and other times when you feel that you will never be creative again. Since Uranus is connected with modern technology, you might find it helpful and productive to put your thoughts into a tape recorder when they are flowing rapidly. You can evaluate and sort them out later. When you are uninspired, do not try to push yourself to be creative. It probably will not work and will only tend to make you restless and nervous. Do something else until the ideas begin to flow again.

You will probably not be patient when it comes to talent and it might be a good idea to avoid artistic endeavors that require a great deal of discipline. For example, if you were to take painting lessons you could find it difficult to painstakingly practice techniques or to conform to the instructor's directions. **Your talents need to flow freely** and you will probably show originality in whatever art form you develop.

Since **your creative talents need not be in standard art forms** with this planetary placement, you may not consider yourself creative. People tend to think that if they do not paint, write, dance, act, play instruments or sing, they have no talent. But, with Uranus in the fifth house you may have computer skills, or a magic touch in fixing cars, or display talents in using other forms of machinery or technology. I have examples of people with this placement that range from telephone salesperson to playing the synthesizer (which of course combines music and technology).

Instead of, or in addition to talents, you may look to this house to explain how you gamble or speculate. Uranus placed there would indicate that you could be spontaneous when such matters are involved. In fact, it is easier for you to take action quickly and back off just as fast, than it is for you to hang on to an investment for an extended period of time. You would probably take risks especially if the potential gain is great enough, but the results would have to be seen soon after you take the initiative. Therefore, your involvement in speculation or gambling may be sporadic, and your successes intermittent. However, the more information you gather (even if it is done a little at a time) before you take action, the less likely you will be to fail.

The fifth house also describes what you need in courtship and, I would add to that, it indicates the way in which you initially interact with others as individuals, whether the involvement is romantic or not. In this area you decide whether a person moves to the seventh house (intimate one-to-one relationships, be they marriage or close friendship), the eleventh house (acquaintances) or no house (out of your life). The sequence might work in this fashion. You go to a party or other group gathering (eleventh house) and find someone whom you deem interesting and want to get to know better (fifth house). Then you can determine if you would like to have that person as a close friend or partner (seventh house), as someone with whom you want to interact personally but not get close to (fifth), as an individual you may only see occasionally at parties (eleventh house), or as someone you never want to see again (no house).

Whether we are investigating potential courtship or friendship, with Uranus in the fifth house, you may be hesitant to form attachments with others quickly because you are concerned about having people interfere with your independence. You could socialize with others but probably in a detached way. You might enjoy intellectual interaction but keep people at arm's length on the emotional level. This, of course, can change when the relationship moves out of the fifth house.

Another possibility is that **you attract, or are attracted to people who are creative, exciting, unusual, unreliable, distant or odd**. These characteristics are not necessarily mutually exclusive. Someone who is stimulating may be eccentric as well. In fact, it could be the eccentricity that leads to the creativity. But if you are really upset about those who are drawn to you, realize that you are probably asking for it. If you can start consciously seeking out those who are creative and exciting, then perhaps the less desirable people will disappear.

Whether or not you are involved with Uranian types, you will probably **crave some excitement when you interact with others**. If you come into contact with an individual who is totally predictable, he or she will undoubtedly also seem boring, and the chances of the two of you forming a close relationship are negligible. However, if you enjoy the company of the other person and think the relationship could lead to marriage or an intimate friendship, you might proceed to the next stage. Then there might be some adjustments to be made. When someone who has been courting you becomes a marital partner, or an acquaintance becomes a close friend, you are moving them from your fifth to your seventh house. And if the requirements of the seventh house are different from those of the fifth, you may wonder why the transition is so difficult. Astrology can help you get over this hump.

By looking at the **signs and planets in the fifth house, you can see how the native interacts initially with others on a one-to-one basis** and also determine the qualities that person is attracted to in others. If you then interpret the seventh house and compare it with the fifth, you can know whether the shift from courtship into marriage or from the getting acquainted stage to intimate friendship takes place easily or requires great adjustment. Awareness of potential difficulties can be the first step in helping to bridge the gap between the two areas.

Since we are dealing with the planet Uranus in the fifth house, let us use this as our example. You have possibly been attracted to the excitement and craziness generated by the other person and decide it would be great to be married to him or her. However, you have Capricorn or Saturn in the seventh house and you need stability and conformity in a partner. It is obvious that problems could arise. But if you are aware that this is eventually what you want in a partner, you can begin to think about it in the early stages of the relationship and weed out those who do not fit the picture before you become too attached to them.

URANUS IN THE SIXTH HOUSE

If you have Uranus in the sixth house you will probably try to avoid a highly structured daily schedule. **You will want variety and challenge** in the tasks to be accomplished. If your job is monotonous or there are too many restrictions placed upon you in the workplace, you could concentrate on changing the routine in a somewhat quiet manner, instigate a revolution, or leave the job. Periodic escapes or frequent expression of your independence may be enough to make an organized office situation tolerable. In fact, you may feel more secure in exhibiting your individuality if there are some guidelines.

Knowing your limitations in the work area could make it easier for you to be creative in your job. You can allow your originality to flow freely if you know that anything that is too outlandish will be curtailed. I have a client with Uranus in the sixth who is a perfect example of this. She complained that her boss did not always give her the latitude she wanted. He would sometimes veto her ideas. When we discussed this further, however, it turned out that he really did allow her to pursue most of her ideas, and only turned down plans that were not feasible or were against established company policies. She even came to the conclusion that her boss's attitude permits her to be more imaginative and creative than she might be otherwise. This is because she doesn't have to waste time thinking about the plausibility of her ideas. It is done for her. But had her boss limited her too much, she probably would not have stayed with the job for the five years she had already been there. Individuals with Uranus in the sixth house would probably quickly leave positions that restrict them too much, or they might try to avoid such jobs entirely.

Uranus in the sixth house could also indicate that **you keep your distance from those with whom you work**, be they bosses, subordinates or colleagues. You need not work in solitude. In fact, you might enjoy the excitement of interaction, especially with those who are intelligent. But you probably maintain a personal detachment, and form your close one-to-one relationships elsewhere.

If you have Uranus in this house, **you do not want other people to tell you how to do your job**, but you probably will not consciously seek out positions of authority either. If you are just allowed to do what you want to do in the manner of your choosing, you should have no problems. Sometimes, however, you could find yourself in a job in which you are managing others. If this happens, you will most likely give your subordinates a lot of latitude, partly to stimulate their creativity, and partly to keep them from interfering with yours. But there could be concern that if your subordinates are given too much freedom they might undermine you or interfere with your manner of operating. It is independence rather than power that is an issue with Uranus in the sixth house. A way to combat that is to periodically get together to report progress and do some brainstorming. The prospect of sharing ideas should be stimulating rather than restrictive. In this way you are kept informed of what is going on, but, since you are only assembling occasionally, you are not weighted down with constant contact.

You might not think about Uranian issues when you apply for your first job, but once you have experienced tyrannical bosses, boredom or other limitations on a job, having a degree of freedom, a bit of excitement, and expressing individuality will become requirements you consciously look for.

You might also discover that **your manner of operating in ordinary life is a little different** from the way in which many other people do. Since the sixth house describes the way you conduct yourself in your daily routine, it can be in the workplace, in the home or anywhere else you happen to be.

The **signs in the house will describe action in the course of the day**. For example, if you have a fire sign on the cusp of the sixth house, you will probably start moving quickly when you get up. Then if you also have an earth sign in the sixth house, following the fire sign, you will probably get organized after you have taken some action. The specific signs will provide more definite information. The planets in that house (if any) will explain your modus operandi even more.

Uranus in the sixth house could indicate that you are not interested in a sedentary daily routine. You might **change your schedule frequently** to keep your interest peaked or just to prove that you are not tied down to a particular routine. But whatever your reasoning, you probably try not to do the same task at the same time two days in a row. Or you might create excitement or controversy in order to avoid boredom. The emphasis with Uranus may be more on action than accomplishment.

You probably do not lack energy. In fact, there are times when you feel "wired" and it is difficult to sit still. But your **energy may come in spurts**; so that you can work feverishly one day or for part of one day, and then need to free yourself of work the next day or part of the same day. Besides the fluctuation of energy level, your attention span could be short as well. Therefore, it will undoubtedly be better for you to switch activities frequently. Your periodic energy overloads can be physical or mental, so you might change the mode of activity as well the specific tasks. In other words, you might perform a mental task first, and when your brain begins to tire, switch to something that requires physical energy until you feel physically fatigued, and so forth. You will accomplish more if you do this rather than trying to remain with one job for long periods of time.

CHAPTER 9

URANUS IN HOUSES 7, 8, 9

URANUS IN THE SEVENTH HOUSE

The seventh house represents both **what you need in close one-to-one relationships and what you have to give in them**. Sometimes we try to project our seventh house requirements solely onto others, and sometimes we look only to ourselves to provide these necessary qualities. But neither of these alternatives is completely satisfactory. You could go from one relationship to another, and from one extreme to the other, until you realize that you must **share** the requirements in that area with a mate, a best friend, or anyone else with whom you are intimately involved.

If you have Uranus in the seventh house and you try to express all the qualities described therein yourself, you will probably insist on **showing your individuality and maintaining your independence in close relationships**. Because there needs to be some stability to maintain a relationship, you will probably attach yourself to someone who is responsible and steady. Also, if you are being nonconforming, your partner will undoubtedly be assigned the task of defining your parameters. Since you are placing the emphasis on responsibility and stability, this person may lack the excitement you need in a partner. Then eventually you could end the relationship because you are bored or feel limited or your partner might terminate it because you are so unreliable.

If the relationship ends, then very possibly your next relationship will be with **someone who is exciting and needs a great deal of freedom**. You will be attracted to the independent nature that was lacking in your last partner. But before long you could discover that this independence brings with it a lack of dependability. So you become the one burdened with responsibility in the relationship. When this occurs, you begin to feel trapped because you no longer have the freedom you need. If the feeling of being restricted becomes very strong, you could abruptly end the association. Then you might go back to a partnership in which you are the independent one and keep alternating these two scenarios.

A possible **third script is to decide not to form any intimate associations**. Having been exposed to both extremes (or even without experiencing either extreme) you may begin to feel that your freedom is more important than any relationship. You believe you will be happiest if you avoid intimacy. You could try to stay away from people altogether, but this will not work for long because you need intellectual stimulation and with Uranus in the seventh house that stimulation comes with and from others. So you interact, find yourself attracted to some exciting, intelligent person with whom you are sharing ideas and start alternating all three scenarios again.

Eventually you might come to the conclusion that you can never have a lasting relationship. Some people with Uranus in the seventh house say that and some astrology textbooks state this as well. It may be self-fulfilling prophecy, but it need not be true. In fact, I can give you a number of examples of people with this placement who have been able to maintain enduring relationships. One couple I know very well have, at this writing, been married for over forty years. He has Uranus in the seventh house and she has Aquarius there. There were adjustments along the way, but they have learned to share the qualities they require in relationships.

Knowing that there are alternatives, understanding what they are, and consciously developing the desirable ones are the secrets to making a relationship work. With Uranus in the seventh house, remember that defining your boundaries is the first step in expressing your freedom. You might make a list (either in your mind or on paper) of what your requirements are in relationships. You can do this with or without astrology, but astrology makes the job easier. Note what you want and do not want in terms of the signs, planets and points in your seventh house, and remember that you and your partner both need to express some facet of these qualities.

You should also be aware that with Uranus in the seventh house, **you probably do not want the kind of relationship your parents had no matter what it was like**. Even if you consciously believed your parents were the ideal couple, you would possibly be attracted to people who were unlike your parents. The differences might in be ethnic, religious and/or socio-economic background. If your parents had a traditional or conservative marriage, you could find the hippie type appealing in partnership, but if they were freethinkers, you might want a partner (at least initially) who is conservative. The principle here is that you will break with tradition whatever it might be.

Let's look at specific issues that could be prevalent with Uranus in the seventh house, and some ways of dealing with them. As we examine the alternatives, it should not be difficult to sort out the desirable and undesirable qualities.

Freedom is one theme that will have prominence in partnerships. As already stated, you might be attracted to the independent nature of your partner (especially if stability was an integral part of your parents' relationship). You do not, however, want a partner who is so independent that there is no need for you. Therefore, as you begin to develop intimacy, look for mutual interests that can be shared, particularly those that entail excitement and mental stimulation. You might join forces in a challenging project or unite behind some cause. However, you do not want to share everything. Twenty-four hour a day togetherness can be limiting. But having some common interests can create a mutual bond to keep you together without making you feel confined.

It can also be satisfying to **find a partner who is spontaneous and exciting**. But with these qualities there may be no thought given to consequences or responsibilities. Unless you establish certain ground rules you could find that all the responsibility is heaped on your shoulders and you feel fettered rather than free. You do not need too many regulations, only a definition of the outer limits. For example, your spouse might walk in and announce that the two of you are going to take a romantic cruise. That could be an exciting prospect, and how nice to have a spouse who would think of such a trip. But if you are left with the details, such as how it will be paid for, who is going to take care of the children, etc., it could take away all the pleasure. If, however, you determine long in advance that any details must be jointly handled when last minute plans are made, you will not feel burdened by the responsibility. And because your partner is sharing the responsibility, you can more easily share the fun.

Another characteristic that can come with the need for freedom is the **desire for periodic time alone**. If you have to be with others constantly, you can feel oppressed, restless and/or exhausted. You want to break away, be by yourself, and replenish your energy (or perhaps your tolerance). Therefore, you want a partner who will at least be understanding of your need for occasional solitude. But chances are that you will be drawn to those who also want some time alone. If you are attracted to someone for other reasons, and then you discover that this person has to be with you constantly, you probably will not stay in the relationship too long.

The two main problems that could arise in connection with the need for solitude are: 1. you and your mate want to be alone at different times; and, 2. concern with what your spouse is doing with his or her free time. If you want to be alone at different times, and do not make an effort to remedy this, you may never get to see each other and then there wouldn't be any relationship. If you are concerned with what your spouse is doing, you could get obsessed with this thought and defeat the purpose of your free time. Therefore, it would be a good idea to jointly schedule times alone (remaining as flexible as possible so that changes can be made), and to share information about your individual activities. You do not want the relationship to become too regimented, but total lack of consideration for the partner (which could occur if you made no adjustment to each other) is just as bad. Once you have developed a loose pattern, which has room to accommodate exceptions, it can become an automatic part of the partnership and you will not feel restricted.

Another possibility for Uranus in the seventh house, is to find **a partner from a different background from your own, either ethnically, psychologically or socio-economically**. You might select such a partner in order to be a rebel, or to emphatically state your individuality. Or you might be attracted to this person because he or she offers qualities that have been lacking in your life. For example, your partner could come from an emotionally demonstrative family, whereas yours was cool and reserved and, through your partner, you might learn to express your feelings more readily. You, in return, might help your partner to exercise more control. Or if your spouse or dear friend is from a foreign country you could expand your knowledge of the language, customs and culture of that country because of the association. And you could share these facets of your own country with the significant other.

The differences between you, however, could also create problems. If each of you is accustomed to behaving in a particular manner that is unlike your partner, conflict can arise. In using one of the illustrations mentioned above, you could handle critical situations in your cool reserved way, by logically evaluating what is going on. While your partner, in his/her typical demonstrative fashion, at least from your point of view, will rant and rave. Or if you are an American "women's libber" married to a man from a Near Eastern country who considers women second class citizens, you can reasonably expect that the two of you will have to make compromises or problems will most assuredly arise.

Compromise is not easy with Uranus, but the objectivity attached to the planet may help to resolve difficulties. As different situations arise, evaluate circumstances and determine together the best course of action. **Discuss the situation objectively**, as though it were happening to two other people. That could help you arrive at a fair compromise.

So **look for a partner who is creative, unusual, intelligent, exciting and/or from a different familial background from your own**. Avoid those who are completely unreliable, totally self-centered and extremely erratic in behavior. You may think that this sounds too difficult, but if you apply your objectivity you will find that it is simpler than you believe. With Uranus in the seventh house you have the ability to view the relationship as though you were not involved. Then, before you allow yourself to be drawn in, you can eliminate those who exhibit a number of the qualities you want to avoid.

URANUS IN THE EIGHTH HOUSE

The natural assumption with Uranus in the eighth house as it pertains to sex is that the native would want total freedom in sexual experiences or would have unusual standards in regards to the sexual theory or practice. And if you have Uranus posited in this house, you may be totally uninhibited in sexual matters. But I have discovered in discussions with clients who have this placement, that there is another possibility which is just as common. You do not want to be controlled in that area. You may enjoy sex, but you want it when, where and with whom you desire it.

One tends to think of passion and lust connected with uninhibited sex. But if you have Uranus in the eighth house, **you can become cool and detached** if you feel that your sexual partner considers you a sex object or tries to control you through sex. You may then make yourself inaccessible even if you do have deep inner passion. On the other hand, if you are respected and feel free in expressing your sensuality, your sex life can be very gratifying.

Since the eighth house is also other people's money and joint resources, you could find that **unearned money comes in sporadically or other people's resources help to make you feel independent**. More often, however, it seems that those with Uranus in the eighth house **want to be financially unencumbered**. You probably are uncomfortable with debt. You may not even want to be responsible for managing funds which are not your own. However, it might not always be practical to accomplish what you want all by yourself.

For example, I have a client with this placement who felt very strongly about owing money. He had always paid cash for whatever he purchased. The one credit card he had, he used only as a monthly charge, so that he would not be in debt to even the credit card company. Then came a time when he wanted to buy a house and the situation was very difficult for him. The house was too expensive for him to pay for out of current earnings. He needed to borrow money from his family for the down payment, and he had to take out a mortgage from the bank as well. I suggested that he view the matter from a different perspective. Instead of looking at his borrowing money as a terrible burden, he should think that each payment he made would make him freer. He borrowed the money. He still has an uncomfortable feeling about the debt, but he is trying to develop the right attitude. He is working hard to pay off his loans and his mortgage and tries to make double payments whenever possible. He told me that this makes him feel twice as free!

Since the eighth house represents joint resources as well as the resources of others, you might also find it difficult to share money or talents with others. If you are married, you could prefer that your partner handle joint money so that you can be free of such burdens. Or perhaps you would rather you each handle your own resources and not pool your assets.

In regard to sharing talents, you might be concerned that too much togetherness could interfere with your creativity. However, if you have the right attitude, you may be able to stimulate the creative process. Brainstorming can be exciting and productive as long as you are not wasting your time dwelling on the possibility of being controlled or thwarted. Always consider this form of sharing temporary, knowing that if the other party seems to be taking charge, you can extricate yourself from the interaction. Thus you may be able to allow ideas to flow more freely.

The eighth house is also connected with psychological probing and transformation. So with Uranus there, you might find that you do not always have the patience to get to the bottom of issues. You can deal with this potential problem by analyzing for a short period of time, taking a break, and returning to it. A number of sessions may be necessary, but insights concerning deep, psychological complexes can occur. The insights will probably come suddenly, and may lead to personal transformation. Total transformation is possible, but it will most likely come as a result of several insights, each of which may occur suddenly.

URANUS IN THE NINTH HOUSE

Uranus in the ninth house can be indicative of someone who strongly believes in **religious freedom for everyone**. Personally, this individual will find it hard to conform to an organized religion. If you have this placement, you might even object to the word "religious" being attached to you. But call it what you will, there is a need for a personal belief system. It is possible to start life in a traditional religion, but unlikely that you will stay in it. You probably consider your religious beliefs unique and it may surprise you when you discover that, although some of your ideas are nonconforming in terms of organized religion, there are other people out in the world who have similar views. This is because the main concern is not necessarily uniqueness or eccentricity, but rather that your spiritual concepts satisfy you as a person—that they contribute to your sense of individuality and make you feel freer.

If you were raised in a traditional religion, you will probably retain some of tenets from your early religious training (although you might do this unconsciously), but you will discard anything with which does not fit into your way of life or does not make sense to you.

A number of people with very different horoscopes, but all of whom have Uranus in the ninth house, have each mentioned to me a **search for a personal religion**. The methods they used varied. Some read books, others took courses, still others visited houses of worship, and some did two or more of the above. However, whatever method was used, what each was doing was to examine the possibilities of the ninth house in terms of religion. They were defining their choices so that they could stretch their boundaries. They wanted **a belief system that would make them feel freer**. The body of beliefs that each ultimately embraced were, to some degree, different from each other, but the overall result was the same—the formation of a personalized religion.

With Uranus in the ninth house there can be an **intellectual curiosity in terms of spiritual matters**. But because Uranus is a fixed planet you are not easily influenced. You do not readily accept ideas without question. However, your curiosity can prod you to investigate and your objectivity help you to look fairly at what you find. You may even examine religious concepts in order to disprove them, but your mind should be open enough to accept tenets that withstand the probing.

Since the ninth house is also the house of higher education, your intellectual curiosity can be satisfied by learning about matters other than spiritual ones. There is a **thirst for knowledge** and an impatience to get it as quickly as possible. If we add the ingredient of impatience to the non-conforming quality associated with Uranus, it is understandable why some individuals with this placement do not easily adjust to college life and may take another route to educate themselves.

On the other hand, many people with Uranus in the ninth house do attend traditional colleges. This may be because the desire for knowledge outweighs the need for freedom, or else they find ways to satisfy individuality and the revolutionary spirit within the frame of reference of the four-year college or university. If you are an idealistic person with a cause, or in search of a cause, you might gravitate toward an institution of higher learning. You may do this because you think that academia caters to **open-mindedness**. Therefore, you will be able to find a cause or get support from other freethinking people. You could consider the campus a good place to vent your revolutionary tendencies.

Another reason for attending a traditional college is that you may take pleasure in **challenging your professors**. Someone I know with this placement told me that when he was in college, he would research topics that were going to be taught in his courses with the idea of discovering facts that might refute anything his instructors would say. He particularly enjoyed doing this with teachers who were dogmatic or those who were ill-prepared. He may not have been popular with the faculty, but a number of the courses in which he participated were probably more interesting and informative because he was there. And he felt that he became more knowledgeable because of the experience.

Another route that might be taken in terms of higher education is to attend a technical school. In such places you can acquire specialized knowledge and can receive training for a specific purpose. This training is often focused on a particular subject and you are not slowed down by having to take courses that seem to have no connection to your goal. For example, you might take a computer course to learn to program computers or to operate them. If you take this route, you **get the information you need as quickly as possible** without the embellishment of seemingly irrelevant courses full of extraneous material. You can apply the results of what you have learned right away and satisfy that impatient quality so prevalent in your attitude about higher education.

Another possibility of Uranus in the ninth house is to avoid any form of traditional higher education and broaden yourself in other ways, such as **travel**. You need not be constantly on the move, but when you go to a different state or country it is not just to vegetate. You want to know what the natives are like and will probably ask about and go to see the points of interest in the area you are visiting. The success of your journey will most likely be measured by the amount of knowledge you have gained.

CHAPTER 10

URANUS IN HOUSES 10, 11, 12

URANUS IN THE TENTH HOUSE

Uranus in the tenth house can mean that **you change jobs frequently**, but there are so many other possibilities that it is a poor excuse for career failure. Although it can indicate a restlessness in the area of occupation, you could also be dissatisfied if you were unemployed for too long. Important requirements are that **you have a degree of freedom** in what you do in the world and that **you are challenged** in that area. If you are not working, you may be free, but you will be missing the challenge.

You are probably most content if you can **use your ingenuity in your career**. You thrive on spontaneous responses to situations and, therefore, prefer not to be involved in a job in which there are no surprises. However, you really do not want a totally unstructured situation, nor do you respond well when you feel another person is using surprise tactics to control you. You tend to resist any change suggested or instituted by someone else. If circumstances demand a new approach, you have the ability to come up with one that is fresh and original. But if you are supposed to carry out someone else's new and creative ideas, you might rebel. And you possibly would seem so unbending that you might be considered a reactionary.

If you have Uranus in the tenth house, your profession should ideally include a broad, stable structure within which you can operate, but enough **flexibility to help you avoid monotony and boredom**. You will want to be aware of company goals so that your direction is clear. But, you prefer to pursue these goals in a manner that you select. And, if left to your own devices, you will probably go beyond the company's expectations.

You could decide that having your own business is essential to your need for freedom, especially if you have had dealings with tyrannical bosses. Working for yourself can be very appealing to your urge for independence. But what you should consider before you embark upon this path, is that there are responsibilities connected with self-employment that are not part of working for someone else. When you have your own business, you have to deal with finances, advertising, marketing and probably a number of other issues, depending on the type of profession involved. In other words, when you are the one in charge, success or failure is based largely on your own efforts or your ability to delegate well.

If you are thinking about starting your own business, you should consider what needs to be done in order to make it succeed. Determine what you want to do yourself and what you would like others to do. You might like the idea of doing everything yourself and, if this is feasible, there should be no problems. If, however, your goals are too large to handle alone, you can still establish a successful business if you analyze your situation and take action accordingly. After you decide what parts you will enjoy doing yourself, find people for the remaining tasks who have capability in the sector of the business involved and are self-reliant, but not so independent that they might overthrow the power figure — you. This last possibility can be avoided if you select subordinates who are trustworthy and also report to you regularly in order to keep you informed of what is occurring.

What you do not want is to feel more tied down in your own business than you were when you worked for someone else. If you have to worry about what your subordinates are doing, you will feel weighted down. Responsibilities may interfere with your creative ideas which you started your own business to express. So if you are feeling uncreative or bored with your career, **do not permit your restlessness to cause you to act hastily**. Before you move in this direction you might first get away from the office for a short period of time. Go out into the world and do something exciting or creative but totally unconnected with your current career. This may be enough to renew your enthusiasm and give you fresh ideas. Another goal or dimension might be added to your business or, possibly, a new type of career could stem from your newfound excitement or creativity.

Another course of action could be to simply **try a new or different approach** to your work. Or you could investigate new equipment available to you that might be exciting to use and possibly streamline your business. That could make your work easier and give you more time to be creative. Also relying on machines instead of people could be appealing to you. A machine may break down, but you don't have to worry about its motivation in doing so. If, after various attempts to put new life into your business, you are still dissatisfied, you might then consider doing something else.

Although Uranus in the tenth house indicates that change can be undertaken in the career, you do not want to make abrupt changes without forethought. Sometimes restlessness and the desire for change, or pressure from others may cause you to act too rashly and you can create a situation that is worse than the one you left. Some of the suggestions mentioned above could be tried, or you might create other appropriate actions. Keep in mind that **a change of attitude may be as satisfying as a change of career**. There is a desire to make drastic changes, but minor adjustments may be all that is needed. If you have made minor adjustments and are still dissatisfied, you might consider a career in a Uranian field such as astrology, computer technology, electrical engineering, or anything unusual or on the cutting edge.

URANUS IN THE
ELEVENTH HOUSE

If you have Uranus in the eleventh house you will probably **avoid attaching yourself closely to a group unless the group is**

unusual. Even then, however, you could find yourself looking for excuses to miss meetings or to break away from the crowd. You are most comfortable on the periphery, viewing the gathering as an observer rather than a participant. This is the easiest way to objectively evaluate the situation and to maintain your independence. You do not want to blend in so much that no one notices you because you do not want to be ignored or taken for granted. But, you may not want to be a continual participant. If you feel ignored, you might do or say something outlandish just to express your individuality.

This attitude may be prevalent in social events, clubs, business meetings or other kinds of gatherings. You could remain **detached** or try to avoid them entirely. You might also find that if you do attend a meeting, or accept an invitation to a party, you have the strong urge to leave soon after you arrive.

There are ways to combat the feelings. You could keep flitting around the room. The movement alone should be comfortable for someone who has Uranus in the eleventh house. But, as you move around, you could seek out people whom you think might be mentally stimulating. If you have no idea how to find interesting people, select someone who dresses oddly or for some other reason seems different. You would probably naturally gravitate toward such people anyway and you may find that you leave the party with new ideas or information, or at least a good story about the strange people you met.

Another way to cope with parties is to **stir up a little excitement**. I don't recommend physical combat, but verbal bantering could be fun. You might express the opposite viewpoint to anything anyone suggests. It should make the gathering interesting and make you feel your attendance was worthwhile. If you evoke strong feelings or hostility in others, a discordant situation might arise and your host or hostess might never invite you to a party again. But you could welcome that response if you didn't want to be there in the first place.

There is still another way you might choose to express your Uranus in the eleventh house. You might **select a cause** with which you could become involved. It is probably best if you do this on a **short term basis**. A long term project could become boring and your interest would undoubtedly wane. Then you either experience burn-out or dissatisfaction with the progress being made. That is the time to drop what you are doing and move on to a new cause. If you have committed yourself for a long period of time, it may not be so easy to leave. Therefore, it is simpler to assume short term obligations in regard to a humanitarian project or cause.

It is possible to stay with a cause indefinitely with the right approach. Make sure that is clear to all parties involved that your commitment is short term and you will not feel trapped. Then, as you are approaching the end of your commitment time, you can review how you feel about what you have been doing. If you think it is time to move on you can do that, or you might decide you want to continue and you can reaffirm the same commitment. Should you decide you want to go on, your evaluation might have instilled a renewed excitement within you, or you might have thought of some creative ideas and an innovative approach to the project. Using this approach, you could discover that you are more of a joiner that you realized.

URANUS IN THE
TWELFTH HOUSE

If you have Uranus in the twelfth house and, as a child, your independence and creativity were encouraged, you will undoubtedly be able to easily express these qualities as an adult. You should have little difficulty exhibiting your individuality and you will probably **easily combine imagination with your creativity**.

If, on the other hand, the independent side of your nature was frequently suppressed, you might have become timid and too conforming. If you are overly cautious and never allow yourself to exhibit the maverick in you, you could stunt your emotional growth. You might spend your life being unhappy about your circumstances, wishing that somehow you could be freer.

If, however, you become extremely restless or dissatisfied, your self-control could disappear and then your behavior might become erratic. This extreme will not be satisfying either. You want to express your individuality and independence, but you have to determine acceptable limits in so doing. Although you do not want to be totally conforming, neither do you want to be a pariah. Therefore, you have to decide how different you can be without being considered very strange or even crazy by others, but it is most important that you approve of yourself.

Another way you might choose to quell the restlessness is to become a "**closet rebel**," nourishing your differentness without fanfare. You could do it quietly, or perhaps deviously. Being covert, however, might give you feelings of guilt. This, in turn, could interfere with the personal gratification you should get from expressing your individuality.

If you have Uranus in the twelfth house and tend to have problems with such issues, you might try the following procedure to get the energies moving. First, probe deeply (twelfth house) into why you feel inhibited. **Stir up old memories** of how your individuality was thwarted and what occurred when you tried to express your independence. This could unlock fears that have been stored up over the years, and allow you to take small steps toward becoming freer and more creative.

You also can consider acceptable outlets for expressing your uniqueness through the twelfth house. You might choose to become involved with **mystical pursuits**. The particular mystical path you select is unimportant because any form of mysticism is considered unusual and sometimes strange by the materialistic society which sets our standards. Delving into metaphysics could enrich your life, and you may begin to feel that freedom you have been missing.

Another possibility is to **become involved in the arts**, particularly music. This will trigger the creative element associated with Uranus that you may not even know you have. Your creativity may be unusual and come in spurts, but demonstrating your uniqueness in this manner can alleviate your restlessness or feelings of suppression. This may seem like an indirect approach to expressing your individuality, but you may find that it is far more satisfying than inciting an insurrection, or being covert about your nonconformity.

As mentioned in the beginning of this section, if your creativity and individuality were encouraged when you were a child, you probably grew up with the ability to openly display your uniqueness confidently, without concern for rejection or repercussions. Your creativity, too, will flow quite easily. Keep this in mind when you are intimately involved with children who have Uranus in the twelfth house. Work on helping to cultivate Uranian themes in their lives.

No matter which house in your horoscope holds your natal Uranus, remember that **this is the area in which you are supposed to break with tradition and also directly display your individuality**. With all the planets, but especially Uranus, the planet of change, you will need to find different ways of expressing the themes in the course of your lifetime. You will know it is time to make changes in terms of Uranus when you feel oppressed or restless. If you recognize these symptoms as soon as they appear, you can immediately define your parameters in the house in which Uranus is placed so that you can begin to break out of the mold quickly. Then you can make the necessary changes with a minimum of effort and a maximum of spontaneity and success.

CHAPTER 11

URANUS IN ASPECT

Thus far we have discussed the meaning of the planet Uranus in signs, and then in houses. There is one more context in which we can view the planet, and that is in aspect. Aspects can be in the natal chart or may be formed by the movement of the planets in the sky after birth. Two of these dynamic techniques are secondary progressions and transits. The significance of a natal aspect may be different from either progressed or transiting aspects. And, the meaning of transiting and progressed aspects may be different from each other as well.

If you have an aspect in your natal chart, you will live with it your entire life. This is a very long time, indeed, especially if you have not found the key for expressing it positively. However, if you look at it from another perspective, you might feel differently. Since the aspect is constantly with you, it can become comfortable —not always good, but nonetheless comfortable. The themes connected with the planets that are joined by aspect usually operate as a unit. You do not have to think about putting them together unless you are unhappy about the way they manifest themselves. Then you **should** become aware of them and consciously make an effort to change the unsatisfactory patterns of behavior associated with the combination. It is highly likely that you will examine and possibly change patterns more than once in your life. This is because there are many ways in which drives associated with the planets can be expressed. And, there will undoubtedly be periods in your life when it will be advantageous for you to manifest one form of behavior and other periods when another expression of the aspect will be more satisfactory to you. Because time and familiarity are on your side, this shift could become a relatively simple task, if you are fully aware of the aspect.

Secondary progressed aspects, as they are forming or becoming exact, may indicate **times to focus on changing patterns**. Progressed planets and aspects can become an integral part of you for the period of time in which they are in range and they gain momentum as they move closer together. Since the concept of secondary progressions is based on the premise that one day in the ephemeris represents a year of life, the outer planets move very little in the course of a lifetime. Ninety days would represent ninety years, and in ninety days the outermost planets (Jupiter through Pluto) will move, at most, only a few degrees.

PLANET	MAXIMUM DAILY MOTION	MAXIMUM DEGREES IN 90 DAYS
Pluto	2 minutes	3°
Neptune	2 minutes	3°
Uranus	4 minutes	5-6°
Saturn	6-8 minutes	10-12°
Jupiter	14 minutes	18°

This means that aspects between the five outermost planets that appear in the birth chart will probably be in range by progression most of your life. If the aspect is exact at the moment of birth, it will begin to separate (becoming less intense) as time goes on. If the faster moving planet is past the slower moving one, it has already begun to separate.

If the faster moving planet is approaching the slower moving one, it is possible that the aspect may become exact during your lifetime and the intensity will heighten as it moves closer. Then it will begin to abate when (and if) it moves past the slower moving planet. Since we are concentrating on Uranus in this book, you should know (and can see from the table above) that Uranus moves more quickly than Pluto and Neptune and more slowly than Saturn and Jupiter.

The remaining planets (Sun through Mars) move more rapidly and, therefore, in ninety days you may see an appreciable difference in their positions.

PLANET	MAXIMUM DAILY MOTION	MAXIMUM DEGREES IN 90 DAYS
Sun	approximately 1°	90°
Moon	11-16°	more than 3 complete cycles
Mercury	2+°	120+°
Venus	1° 16 minutes	90°-114°
Mars	48 minutes	60°-72°

Even though Uranus moves very slowly, aspects which are formed with the five inner planets at birth might move out of orb by progression, or may separate from an aspect and then form a new aspect. This is always true of the Moon. Your natal chart always describes your basic character. **Progressions may accentuate certain qualities when aspects within your birth chart become exact, or indicate temporary conditions when aspects are formed by progression that do not appear in a natal chart**. A year or month in which an aspect becomes exact by progression will be an important time for that combination, whether it involves a progressed planet moving to a progressed planet or to a natal planet.

Aspects formed by transiting planets can be quite different from progressed aspects. Unlike progressions, **transiting planets** do not usually become an integral part of us. They indicate energies that are available to us at particular times and seem to **place a spotlight on the planets in the natal chart they are activating**. So the natal planet informs you of what you should be working on within yourself, and the transiting planet tells you how. For example, if transiting Neptune is aspecting your natal Saturn, it is time to dissolve a structure in some way, but if transiting Saturn is aspecting your natal Neptune it is time to clarify something that has been vague.

Aspects from transiting planets to natal planets are usually in range for a shorter period of time than progressed aspects. Instead of one day in the ephemeris representing a year in the life, one day represents one day. So, although Uranus only moves about 5-6° in 90 days, by transit it will cover an entire sign in about 7 years, and complete a 360° cycle in around 84 years. It is still moving slowly, but not as slowly as by progression. The five outermost planets move as follows:

PLANET	DURATION OF CYCLE	MAXIMUM DAILY MOVEMENT
Pluto	248.40 years	2 minutes
Neptune	164.79 years	2 minutes
Uranus	84.02 years	4 minutes
Saturn	29.46 years	8 minutes
Jupiter	11.86 years	14 minutes

The movement of the Sun, Moon, Mercury, Venus and Mars is as follows:

PLANET	DURATION OF CYCLE	MAXIMUM DAILY MOVEMENT
Sun	1.00 year	1°
Moon	27.0+ days	15+°
Mercury	1.00 year*	2+°
Venus	1.00 year**	1°16'
Mars	1.88 years	48'

* Although Mercury has a cycle of 87.97 days, from the perspective of the Earth it is never more than 28° from the Sun, and therefore appears to have a cycle of about the same duration as the Sun.

** Venus has a cycle of 224.7 days, but from the perspective of the Earth it is never more than 48° from the Sun, and therefore appears to have a cycle of about the same duration as the Sun.

Since the five outermost planets move more slowly than the Sun, Moon, Mercury, Venus and Mars, it logically follows that their impact would appear to be stronger and their duration longer than the five innermost planets. In dealing with transiting aspects, it is standard procedure to use a 1° orb approaching and a 1° orb separating. However, with Jupiter, Saturn, Uranus, Neptune and Pluto, their significance is usually felt for a much longer period of time than indicated by the 1° orb. Frequently, these planets can move several degrees away from a natal planet, go retrograde, pass over the natal planet a second time, go direct and pass the natal planet a third time. When Pluto is moving at its slowest speed (31 years to transit a sign), it can aspect a natal planet as many as five times. What this seems to indicate is that you have at least three chances to use the aspect beneficially. Actually you have more opportunities than that because the aspect can be evident periodically for the entire period that the transiting planet moves back and forth over the natal planet even when it is more than 1° away from the natal planet.

In interpreting the aspects **it is impossible to tell just from the aspect involved whether the native is using the inner drives positively or negatively**. In traditional astrology, the student is taught that certain aspects are always wonderful and other aspects are always bad. In the real world, however, this simply is not true. Just because a trine flows easily does not mean that it is always used propitiously, nor does an opposition indicate that you will always feel pulled apart in terms of the two planets involved.

Since any aspect can be good or bad, there will be no distinction made among them in the following chapters on aspects. It is the combination of the planets that will be emphasized. Their definitions will be the same no matter what aspect is involved. It may be that what aspects describe is the best manner in which to deal with the planets connected rather than the end result of the connection. Therefore, you might want to consider the definitions of the aspects given below and use them as an approach to bring out the best of the meanings ascribed to the planetary combinations in the following chapters. It will be up to the reader to determine what is applicable.

The **conjunction** is the strongest aspect. It is considered a neutral aspect because supposedly some planets fit together better than others and thus, in traditional astrology, some combinations are considered innately good and others innately bad. For example, the "greater and lesser malefics" (Saturn and Mars) have to be bad in conjunction, and the "greater and lesser benefics" (Jupiter and Venus) have to be good in conjunction. You need only look at how certain individuals with these conjunctions operate to know that this is not true. It may be that some people with Saturn conjunct Mars do not accomplish a great deal, but we could find at least as many with the conjunction who are disciplined in taking the initiative and achieve success. It is also true that those who have Jupiter conjunct Venus may be charming and pleasant, but such people may, instead, lack self-control and be overly self-indulgent.

What we can be certain of is that the **planets involved in the conjunction are visibly connected in the natal chart**. Since they are easily seen, it logically follows that perhaps they should be dealt with directly rather than subtly. This means if the planets are not synchronized, you can try overt ways to bring them together, rather than deeply analyzing and realigning yourself inwardly.

The **opposition indicates a need to balance** inner urges. You could feel with this aspect that you are on a seesaw, and without effort find yourself accentuating one side of the opposition more than the other. Often you become aware of this when you feel that something is missing in your life. You can then focus on the other end of the opposition and tend to ignore the energies you had been expressing so well. If we add an image of a small child alone, trying to ride the previously mentioned seesaw, the concept of the opposition can become very clear. Picture this child first sitting on one end, then, when no movement occurs, running to the other end, and continuing to run back and forth until exhaustion stops the process. You will have an idea of the negative side of an opposition. So this behavior pattern can continue with an opposition until you give up with exhaustion and decide you just have to suffer, or you become aware that you need to have balance between the two.

Awareness is a keyword for the opposition. The child in our example may discover, through experimenting or asking questions, that if another person is on the other end of the seesaw, the child can have a nice ride. This should give you a clue as to how you can begin to balance the energies symbolized by the planets involved in your opposition. First, keep in mind that the energies need to be balanced, not blended. There will be times when it will be more appropriate to utilize one side of the opposition more fully while the other side is in the background, and vice versa. But never should either side be totally ignored. When a situation arises that could involve your opposition, fully investigate the circumstances from the perspective of both planets. Determine possible ways to utilize the energies and experiment with them until you find that balance that will make you feel comfortable. This will not be a procedure that you perform only once in a lifetime. Circumstances change so that certain behavior may be appropriate for one situation and not another. Once you know the approach, however, you can use it more and more readily.

Trines flow easily. The planets involved automatically blend together. **They usually perform as a unit without effort or even thought**. You might not be aware that they operate together until someone points out how well you use them, or you find yourself in a difficult predicament because you were carried away. When trines are operating in a manner that is beneficial for you, you need not give them any thought. And chances are that you will not. However, should you become aware that you feel out of control or overwhelmed, you might look to the trines in your chart. Try to redirect the energies. You may not be used to consciously dealing with the trine and, at first, you could feel as though you are swimming upstream. However, as you manifest new qualities associated with the planets, they will gain momentum and move as quickly as they did before, but in a more desirable direction.

Sextiles, as trines, seem to flow easily and we may not always be aware of the planetary motifs **blending** unless we are having problems with the themes connected with them. Sextiles are considered to be help from others (whether you want it or not). And, usually you become aware of them when someone is helping you in a manner you prefer not to be helped.

You may feel like the little old lady who was helped across the street by the Boy Scout. Her problem was that she didn't want to cross the street. She could have mustered all her strength and gone back on her own to the other side of the street, or found someone else to help her. The latter solution would probably have been the easier. With the sextile, it will usually be more efficient to look outside yourself for clues to redirecting the energies, rather than try to solve matters from within. With both the little old lady and you, there will undoubtedly be someone around who can help.

Squares are considered aspects of **manifestation**. Unlike trines, which may be taken for granted, squares are evident because something between the planets involved has to be reconciled. They do not easily blend together. When one planet is activated, the other seems to advise you that something is not right about the first planet. It is therefore hard to ignore the square and it usually triggers you to take action. What you decide to do with it will determine whether (as so many astrologers say) it is a building block or a stumbling block. Squares may take effort, but they can help to build character and lead to achievement.

That covers the Ptolemaic (major) aspects, but to complete the 30°-multiple aspects, let us briefly discuss the **semi-sextile** and the **quincunx**. These are both aspects of adjustment. There may seem to be friction between the planets involved and, when they are activated, you may feel that you have to shift gears frequently. By realigning the energies, dramatic changes can occur.

There are, of course, other aspects you may want to use. You might want to include the **semi-square** (half a square) and the **sesqui-quadrate** (a square and a half) which have the same general quality as the square although not quite so pronounced. If there are other aspects that you use frequently, you can add those yourself. (Squares, oppositions, semi-squares, and sesqui-quadrates are also called "hard" aspects. Trines and sextiles are referred to as "soft" aspects. Any aspect may be good or bad, but hard aspects usually require more effort than soft aspects.)

As each planet is covered in connection with Uranus, there will be distinctions made between natal, progressed and transits when the emphasis is different. As already stated, the particular aspects will not be mentioned. It will be up to the reader to apply them if and when it is necessary.

CHAPTER 12

ASPECTS OF URANUS TO THE SUN, MOON, MERCURY, VENUS AND MARS

Uranus and the Sun

The Sun represents the visible part of us and, when Uranus is connected with it in the natal chart, you will probably be seen as a **high energy person, an individualist, and possibly even a bit of a revolutionary**. There can be an air of **excitement** that surrounds you so that it is not easy for you to fade into the background. This could be modified to some degree by the sign(s) and house placement(s) involved, but there will always be some electricity around you.

Because the Sun also describes our ego needs, Uranus combined with it can indicate that there is a **changeability in what you need for ego gratification**. You may periodically change your goals or direction, or perhaps the challenge and excitement of reaching for goals is more gratifying than attaining them. Once you have accomplished something, you could feel that it is time to move on.

There are other alternatives, however. I have a client with Uranus conjunct the Sun in the tenth house who developed a pattern that epitomized this combination. He was an entrepreneur who would start a business, run it for a short period of time, turn it over to someone else and go on to a new one. For many years he was satisfied with his career. He enjoyed his work and was successful at it. But he began to feel that he had accomplished very little. He saw other people seemingly content with what they had attained, and he was tired of never feeling satisfied with what he had. He no longer wanted to flit from one business to another, but he was concerned that he would become bored if he stayed in one. Using the principles of Uranus and the Sun, we worked out a plan for him to try.

Since Uranus tells you that you want to break with tradition, when it is in aspect to natal Sun **you may choose unusual goals that can bring you ego gratification**. That could help, but the feeling of accomplishment and self-satisfaction might not last for too long. The changeability and restlessness associated with Uranus could trigger the urge to move on. As long as it is satisfying to go on to another goal, you can keep doing this.

My client had been doing it for many years. But, he became dissatisfied with the pattern and, therefore, it was time to develop a new approach. Instead of going in a totally new direction, I suggested that he stay with the current business, examine what he had attained thus far and determine how he could stretch his boundaries (which is another acceptable definition of Uranus). He could continue to search for new challenges, but use these to continue building on what he has already achieved. So, instead of feeling inconsistent and scattered, he might feel more focused.

My client liked the idea, and decided it was worth trying. He reports back periodically and, so far, seems content to continue on that path. There are days when he feels trapped or bored, but he has combated this by taking a little time off to do something that will make him feel free, independent or creative. There may come a time when he will want to renew his old pattern or go off in a totally different direction, but for now, his present path is gratifying.

Besides visibility and ego gratification, there is still another way to interpret the Sun. The Sun can also represent the father. In your natal chart it describes your experience of your father (or the father image in your life). It may not indicate how the world sees your father, or even how your siblings view him, but you should be able to relate to his description in your chart.

When the Sun is aspected by Uranus there are a number of scenarios which might apply. One is that **your father might have literally left you** or, if he was around, was aloof. Or he may just have **been different**. He could have been an individualist who was ahead of his time which might make him **exciting and creative, or eccentric and strange**. You might have been proud of his inventiveness or embarrassed by his differentness. Another possibility is that he could have had an impact on your creativity by stimulating you or thwarting you. **He might also have had an effect on your independence**. Because you have Uranus connected with the Sun, the need for independence can be an important part of your character whether or not your father influenced this facet of you. But it is possible that your father stimulated you to become independent. This may have occurred by his actively helping you to be yourself, or by ignoring you or abandoning you so that you were forced to become independent.

By interpreting your relationship with your father in light of the Sun/Uranus aspect in your chart, you may better understand the influence he had on you. This might, in turn, help to explain why you are as you are. However, ultimately the aspect is yours and even if you had a disastrous relationship with your father, you need not spend your life suffering from it. Look for ways to express your freedom and originality to attain recognition. Find goals that are challenging but can be achieved in a short period of time, and give yourself permission to move on after you have accomplished your goal. With the right approach, this aspect can be an asset and you may be viewed as an exciting and successful individual rather than an irresponsible eccentric.

No matter how hard you try to work with any aspect in your natal chart, there will undoubtedly be times when you will be content with the combination and other times when you will not. For the most part, you learn to live with the aspect, but there are many choices as to how you can express these inner drives and there will be times when you become dissatisfied with the way they are manifesting in your life, as was the case with the aforementioned client. This informs you that it is time to make changes. The timing can often be explained astrologically by progressions and/or transits along with clues as to how this can be done.

Progressions can be interpreted in two ways. They can be treated somewhat like transits, offering information about what is happening to you externally. But progressions, as mentioned previously, become part of our character so they describe psychological and internal developments. Even if they are connected with external events, there is a deep impact on the personality.

If you were born with a Uranus/Sun aspect in your natal chart, you will always live with that combination; but if the aspect is not exact and the Sun, which is the faster moving planet, is behind Uranus in the zodiac, the connection will become closer by progression. Since the Sun moves about 1° a day and a day in the ephemeris represents a year of life, it will occur fairly early in life. If you use a maximum of 8° for a natal aspect, the Sun and Uranus would have to be within 8° of each other to be in aspect and, therefore, the aspect would have to become exact by progression by the age of 8. So matters connected with freedom and ego identity will probably be in the foreground at the time the Sun by progression exactly aspects Uranus, and possibly one year before to one year after (one degree approaching and one separating).

Progressions become part of you and, at the time the aspect became exact, you might have become more of a rebel. Or perhaps you felt that your individuality was threatened. But however it may have manifested itself at that time in your life, it is helpful to know about this time frame. If problems arise in adulthood regarding the themes connected with this combination, you know the time frame on which to focus to help explain why you have these problems and, possibly, clues as to how you might deal with them in the present.

If there was not an aspect between the Sun and Uranus at birth, and the Sun by progression moves into aspect with Uranus, you will probably find, for the period of time it is in range, that you will temporarily become a rebel, feel very restless or strive for more independence or unusual goals. You will probably be more daring and, therefore, more willing to take risks than previously. It may also be a period of change. You might later refer to this period in your life as your "crazy period."

If Uranus and the Sun are aspecting each other in your natal chart and Uranus is at an earlier degree in the zodiac, it may progress to an exact aspect to natal Sun during your lifetime. If this occurs, the definition would be the same as that given above. The main difference is that changes may seem to appear suddenly, but because Uranus moves so much more slowly than the Sun, it would be inching its way to exactness. Therefore, it could begin to be felt a number of years before it was evident to the outer world. It would also take much longer to separate. Thus it would be more difficult to establish the exact time frame of your "crazy period."

Transiting planets move much more rapidly than progressed planets. With transiting Uranus, the maximum movement is 4 minutes a day so that it moves through a sign in 7 years, whereas progressed Uranus moves only about 6° in an entire lifetime. Just considering the difference in the rate of movement between the transiting and progressed Uranus would indicate that the suddenness associated with the planet would more likely be experienced with the transit than with the progression.

This is due not only to the speed of the planet but also because transits bring external energies into our lives. They do not become part of us. They merely spotlight the planets and points in our natal charts by forming aspects with them and advise us to concentrate on actions proposed by the natal planet. The transiting outer planets are usually more important than the inner planets because they are in range for quite a while. Because they linger, they can be associated with certain feelings. So it is common for those experiencing Uranus by transit aspecting their Suns to feel an **unanticipated restlessness**, and just as quickly find themselves dissatisfied with their accomplishments. There can also be a **sudden urge to make drastic changes in life direction**, although the specific direction may not be clear.

If you do not have a connection between Uranus and the Sun in your natal chart, transiting Uranus activating the natal Sun might bring out feelings or ideas that are alien to you. Being prepared for what they might be could keep you from being overwhelmed when they arrive. You could go beyond this, however. You could have a plan of action ready for when the transit occurs, or start such a plan of action even before transiting Uranus moves into the picture.

You might look for stimulation through a new and challenging endeavor. In fact, since the restlessness might materialize as impatience, perhaps you should try more than one new and challenging endeavor. If you can keep your tasks short-term you can see results quickly and receive personal recognition sooner. Then the restlessness and dissatisfaction could be replaced by excitement and anticipation of more of the same.

At the times that transiting Uranus aspects your Sun, you might also be feeling a need for personal freedom, and/or have a strong desire to be noticed by the outer world. If you feel ignored or unappreciated, **you could find yourself taking uncharacteristic actions just to be noticed.** You probably will be noticed because the Sun is the visible part of us. Keep this in mind, because, if your behavior is too bizarre, you may not attain the kind of attention you want.

First, define what is bothering you. Next, determine what goals you would like to achieve. Then take the appropriate action to move in that direction. If you follow this procedure when transiting Uranus leaves the scene just as rapidly as it arrived, you will be less likely to find yourself in an untenable situation.

Whether or not you have the patience to follow the procedure, one thing you might do is to **find small ways to satisfy the Uranian urges for freedom, individuality, independence and creativity**. Instead of packing your bags and taking off for parts unknown, you might just sneak off by yourself for a few hours. Then perhaps your real life situation might not seem so oppressing when you return. One client going through a number of Uranus transits found that she didn't even have to leave home to relieve the restlessness. She rented adventure videos (both adventure and videos are ruled by Uranus). Then when she had the urge to flee or do something her more practical side told her she might regret, she would watch the videos and found they calmed her restlessness.

If you want respect along with recognition when transiting Uranus is aspecting your Sun, you probably will not shave your head or walk into a party standing on your hands. You might instead, try a new hairdo that is very different from your usual style or find a way to enter a party that would be notable, but not weird. Just changing patterns is what you want to do, and there are a great many things to try.

When the transiting Sun is aspecting your natal Uranus, the impact is not likely to be as strong as when transiting Uranus is activating your Sun. The main reason for this is that, using a 1° orb, the aspect will be in range for a maximum of three days. Because the Sun moves so quickly, you may not have time to worry about the implications of the combination. However, you could still make use of it. Since transiting Sun aspecting natal Uranus indicates that the spotlight will be shining on your individuality, creativity and independence, you might select such time frames to express any or all of these qualities.

Uranus and the Moon

Uranus aspecting the Moon in a natal chart could indicate that **you express your emotions easily, but most likely in spurts rather than a steady stream**. The extremes of this connection could be total emotional freedom, to erratic mood swings, or rationalizing away feelings, with a gamut of possibilities in between.

It can be beneficial to release your feelings. If you hold them in for too long you could become bitter and even make yourself physically ill. Or you get to a point where they come out even if you wish to contain them. None of these alternatives is satisfactory, nor necessary. Illness is certainly not a solution. And periodic emotional explosions could make you feel out of control. You would undoubtedly agree that you want emotional freedom, but loss of control is not the same thing. Others could take advantage of this tendency by learning how to "push your buttons." So every time the right buttons are pushed, you may respond emotionally, but it is on someone else's terms. Therefore, you will not experience the **emotional freedom** you want.

To use this aspect to your best advantage, you must first accept the fact that you should not (and probably cannot) hold in your feelings forever. Becoming ill when emotions are aroused may relieve the emotional stress, but the physical stress that replaces it can hardly be considered a satisfactory substitute. The other extreme of holding back your feelings until you explode, will also not make you feel better. You might hold back your feelings so long that when the outburst occurs, no one (often including you) understands why it is occurring. Then you could appear and feel emotionally unstable—which is a far cry from emotional independence.

It is better for you, personally, and for your image in the world to acknowledge the fact that you should **express your feelings** and give yourself permission to do this. Allow yourself to be spontaneous. Let your emotions flow out as you are experiencing them. For instance, at the first spark of anger—show it. If you do this as soon as the anger arises, it is easier to let go of it entirely. If you hold the anger in, it will probably build and lead to explosions. Also, expressing the anger after it has become in-grained will usually not totally remove the feelings. So there may be a second explosion over the same issues and possibly a third, or even more. Whether it is anger, love, fear, or any other emotion, as you allow your feelings the freedom of expression they need, you will be more consistent in your behavior and feel more stable, as well as being viewed by others in that way.

It may take effort to break the pattern of holding in and erupting, but you need not do it all at once. Try it on small, unimportant matters. It may initially seem to you (and possibly others) that you are overreacting, but it may lead to your being able to more readily express your feelings on bigger issues.

If the direct approach seems too difficult, **do something that you find stimulating**. If you are feeling angry or depressed you may not be inclined to redirect your energy. It certainly is easier to stay angry or to wallow in self-pity, but it will be worth the effort because, until then, there can be no improvement in your feelings. By taking some action, no matter how small, you may at least temporarily take your mind away from feelings of anger or fear or sorrow, and it may even do more. A surge of adrenaline could replace the negative emotions with feelings of excitement.

Aside from emotions, the Moon represents the nurturing part of an individual's character, along with attitudes about home and family. In aspect to Uranus, it could indicate that nurturing is probably intermittent, and that you might have to deal with issues of attachment to home and family versus your own need for personal independence. If you are not aware of the seeming inconsistencies, you can begin to wonder about yourself and your ability to settle down and take care of others. As with any aspect, awareness is the first step in handling the situation. Once you understand this facet of yourself, you can work out ways to deal with it. One solution is to set aside some time to explore and feel free. Make this part of your schedule so that the people who are important to you will not think that you are trying to escape from them. By allowing yourself periodically to do your own thing, you will be better able to fulfill your needs to nurture and belong when you return.

The Moon not only describes your own mothering instincts, it also provides information about how you experience your mother, which in turn might explain why you express your emotions in the manner you do. **Uranus/Moon could be the mother who was independent and did not have the time nor the inclination to play the mothering role, or was inconsistent in that role**. Sometimes she may have been supportive of your feelings and encouraged you to express them, and at other times insisted that you control them. This kind of conduct could certainly lead to erratic behavior on your part. The culprit need not have been your mother. Anyone significant in your early life could have triggered the pattern. Understanding how the pattern developed can help to combat it.

Another feeling that I find clients having this aspect sometimes have to face in regard to the mother is **embarrassment**. Your mother may have been neither independent nor unmotherly. Perhaps it was evident that she cared a great deal about you and you did not question her love. She may just have been different from your friends' mothers and you may have found this embarrassing. As children we want to be accepted by our peers even more than we do as adults. So aside from embarrassment, you may have begun to repress your emotions so that others would not notice you or your mother. Then neither of you could be considered different. This pattern of repression then continues into adulthood and has to be recognized before you can begin to change it.

There is also the possibility with a Moon/Uranus connection, that you experienced your mother as a wonderful person. **Perhaps she was creative and encouraged your individuality and emotional expression** or you might have regarded Mom as a friend or peer. Then as an adult you might not have difficulty with the Uranus/Moon aspect. If any of the above is true, you may often make good use of the aspect by being spontaneous in expression of feelings, and also manifesting the creative side of the aspect. You may still sometimes hold back your emotions for too long. Don't be too upset if you occasionally exhibit erratic behavior. That is, after all, one of the definitions of Uranus/Moon. If you dwell on the negative, you may reinforce it. Instead, it may be better to accept that it has happened and take some action in the opposite direction. With each positive expression of the aspect, you might find the negative slipping more and more into the background.

If you were not born with the aspect, as the Moon moves by progression, it will frequently aspect your natal Uranus. Since the Moon completes a cycle in 27+ days, it will form every conceivable aspect with your Uranus in a 27+ day period by transit, and a 27+ year period by progression. Whether the aspect is hard or soft, you might find yourself a little more emotionally inconsistent during those periods and a little more emotionally demonstrative. It is hard to measure the exact duration of an aspect. If we use 1° approaching and 1° separating for the progression, you will feel the aspect for about three months, but it may be a little longer or a little shorter. Your inconsistency may be constant or your **emotions may erupt sporadically**. It is important to recognize the symptoms and do something about them. If you were not born with Uranus and Moon aspecting each other, the feelings associated with this combination will seem alien to you and, when you are having this aspect, you may find yourself worrying about your emotional stability.

Although an aspect occurs fairly frequently in a 27-year period, the intervals may be too far apart for you to remember the last time you experienced the same emotional pattern, and you could be overwhelmed and panicked by your feelings. Now that you have read this description, however, you can look ahead to see when another aspect from progressed Moon to natal Uranus will occur and prepare for it. Or, even if you do not plan any course of action, you at least will not be too surprised when it arrives.

If you behave erratically or have emotional outbursts during the time the progressed Moon moves into aspect with natal Uranus, there may be just cause for it. It need not come totally from within for no apparent reason. Since progressions may be either internal or external, sometimes the timing of the progressed Moon moving into aspect with natal Uranus coincides with emotional upsets in your life. Perhaps people who are close to you are expressing their urge for freedom, or are acting strangely, and you have to cope with their behavior rather than your own.

Whatever the "reasons" that trigger your actions or reactions, you have to start from within to deal with the feelings. You cannot control what others do, but you should have some say in the way you act. If someone else is acting out your Moon/Uranus aspect, you might see very clearly what is happening, but trying to cope rationally with the situation could be a waste of time. You might first want to try to point out to the individual who is being irrational what is happening in practical terms, and how he or she should behave. And it may work. But you may find that what makes practical sense has no impact whatsoever on the other person. This, then, can lead to more internal frustration and external outbursts for you.

It makes sense that if you are coping with feelings that seem to have no connection with your external conditions you would have to work on yourself. But even if you believe that situations which are forced upon you create your instability, it is easier to begin from within rather than try to change your external circumstances. If you concentrate on your own behavior it may be that your environment will improve, or that the conditions that upset you so much will become unimportant. Therefore, you should **find ways in which you can feel more independent in general, and specifically how you can express your feelings more openly in a comfortable manner**.

Since your progressed Moon will aspect your natal Uranus for only a few months at a time, you only have to find temporary outlets. You might get involved, for a short period of time, in a cause in which your ability for venting your emotions will be considered an asset. Your behavior could stimulate others and contribute to the success of the project. You may not want to make this kind of participation a lifelong commitment, especially if it is only a temporary emotional state you are experiencing. In fact, if you are basically a calm, collected person, when the aspect leaves, you will probably find it exhausting (if not impossible) to maintain such a high level of emotional excitement. But a venture of this kind can be an appropriate vehicle for restoring your equilibrium. Because you have such an outlet, the other segments of your life might remain or become relatively calm. Also, if your emotional insecurity has been aroused by events in other areas of your life, your success with this cause could help you to feel better about yourself.

Causes are not the only acceptable choice you have. There are many places and activities where emotional expression is encouraged. Theaters are places where you are expected to show your feelings. No one is shocked or offended by someone crying at a sad movie or play, or expressing anger when the villain has performed some dastardly deed.

Although the Moon is not usually connected with physical action, Uranus can be because it is associated with nervous energy. Therefore, potentially excessive emotional displays might be curtailed by physical activity because this can burn up the temporary extra energy. Although physical activity makes us or keeps us all healthier, exercise based on Uranus aspects will probably not be done on a regular basis. If you were to join a gym because you had a Uranus aspect, with the expectation of going there every day or even three definite days each week, you probably will be disappointed. Obstacles might get in the way of your going. Or when the appointed time arrives, you just do not feel like going. With Uranus it is better to be spontaneous about physical activity. When you feel the anger raising your blood pressure or emotions about to bubble over, you might take a brisk walk, go for a bicycle ride, or find some other way to dissipate the energy.

We have not yet discussed progressed Uranus aspecting natal Moon. The reason for this is that Uranus is so slow-moving that if you were not born with natal Uranus aspecting natal Moon, you will probably never experience progressed Uranus aspecting natal Moon. If you were born with the aspect, the definition of the natal combination would be applicable all of your life so you would not need temporary ways of dealing with the aspect. If the aspect becomes closer during your lifetime, and especially if it becomes exact, the intensity of the aspect might heighten. You would always be living with the aspect but, as it becomes more exact, you would undoubtedly become more consciously aware of it. You might just begin to focus on the qualities associated with the planets when events occur that bring them to the surface. Therefore, you need not develop elaborate plans to work on understanding or utilizing the combination.

You can, however, apply any of the preceding suggestions for the progressed Moon aspecting natal Uranus to an aspect of transiting Uranus to natal Moon. Emotional ups and downs, or sudden changes in feelings are very common with both progressed Moon aspecting natal Uranus and transiting Uranus aspecting natal Moon. The main difference is that when you are experiencing the progressed Moon aspecting natal Uranus, there is the possibility that you will be aware that your feelings are fragile and be cognizant that they can erupt at any time. Even though the condition may be temporary, you know that it is part of you while it is in range. Whereas with transiting Uranus aspecting natal Moon, eruptions may come suddenly and unexpectedly, and are usually triggered from outside of you. They do not become part of your character as progressions do.

A client confirmed this for me recently as transiting Uranus began to aspect her natal Moon. She told me that she never cries, but lately she has become very sensitive and remarks from others that ordinarily would not bother her cause her to cry. She went on to say, "This isn't me and I am very embarrassed by it. It makes me want to avoid people." The difficulty is that she has to deal with people daily on her job. So I suggested that she try watching movies or reading books that are sad or in other ways arouse the emotions and she is finding this helpful. Knowing that there is a reason for the state she is in, and that it is temporary, has also relieved her concern.

Usually with transiting Uranus aspecting your Moon, there will be external events that trigger erratic behavior. You may lash out emotionally or internally experience emotional restlessness and changeable feelings that are not overtly expressed. You may have unpredictable mood swings, or simply find that your response to particular situations is inconsistent. One day you could be satisfied with your circumstances, and the next day wonder how you could have felt that way the day before. The best way to deal with this is to **release the feelings but avoid making drastic changes quickly**. If you move too far too rapidly, when your feelings change, you may find yourself in a position that is not to your liking. And it may be difficult, if not impossible, to return to prior conditions. Instead find someone—a therapist or a nonjudgmental friend—to listen to your ravings. This serves several purposes. First of all, you are expressing your feelings, not bottling them up and this may alleviate your emotions so that no changes need be made. Secondly, it will slow you down enough to consider alternatives before you take action. And finally, the person with whom you are sharing these feelings might give you insight or support.

Because of the **inconsistency** of this combination, it is fairly common for people experiencing transiting Uranus aspecting natal Moon to occasionally question their sanity. Just knowing that it is part of the aspect and that it is a temporary condition, as with the client mentioned above, can help. Then you can concentrate on making use of the creative elements of the combination and take small actions to satisfy the desire for excitement and freedom.

Since the transiting Moon moves about 1° every two hours its aspects to natal Uranus are only in orb for about four hours. Although the aspects may coincide with periods of sudden emotionality, a four-hour period of such feelings would probably not cause concern. I do not usually suggest running away from difficult situations, but when you are dealing with only a four-hour period, you might choose to stay away from people who tend to push your emotional buttons for that short period of time. Or better yet, you might select this time frame to express your emotions.

Ordinarily I do not interpret aspects from the transiting Moon to natal planets unless a client is going through an emotional period, or a particularly emotional day. For example, I did lunar transits for the day one client was going to court to obtain a divorce decree. I explained that emotional events might occur during particular four hour periods, but they would be only in range for that period of time. She reported back that it helped to relax a little more knowing that events would be short-term (as was the case when her lawyer showed up an hour late while transiting Moon was squaring her Neptune) and she was able to make it through the day.

Uranus and Mercury

Natal Uranus aspecting natal Mercury can be indicative of a **quick and creative mind** with the ability to **rapidly grasp and process information**. If you have this combination in your natal chart, you probably also respond quickly to ideas and are rarely at a loss for words.

There can, however, be the tendency to be **impatient when learning new material**, especially when it is long or complex. Skimming may be more comfortable than painstakingly following every word. Therefore, if you are dealing with complicated material, you might miss important details. So you should not try to absorb complex information in one sitting. If you start by skimming, you can get the general idea and then go over it a piece at a time to fully grasp it. It is better to break up the learning period into small time slots. If you find your mind wandering when you are supposed to be studying, you might make the study periods shorter. If you expand your mind in small doses, you will still be able to show your brightness and you will also probably retain the information longer.

Although you may have both **mental and verbal agility**, there may be times when your mind moves faster than your mouth and you could tend to **stutter or to skip over words** that might be crucial to what you are saying. Should you find yourself stuttering, just stop talking and try to express one idea at a time. If you omit words, the people with whom you are trying to communicate will let you know, and you can repeat whatever they have not understood.

Because you think quickly, you might not have the patience to deal with people who are slow talkers or thinkers, or are more cautious in their communications than you. You could tune them out and not hear what they have to say. Or you might finish their sentences for them and your anticipation of their words may not always be correct.

Although **you may sometimes think faster than you talk, there may also be times when your mouth moves faster than your brain**. This might cause others to view you as refreshingly honest (especially if they are not the subject of your honesty), or tactless (if they are). Because you are not weighing your words before you speak, you might later regret what you have said. Another possibility connected with your speaking before thinking is that when you are dealing with people who are quiet or slow in expressing themselves verbally, you might feel that you have to keep talking and afterwards feel that you made stupid remarks. A simple way to handle this potential problem is to avoid, as much as possible, individuals who are very slow or too cautious in communicating. If you are forced to interact with them, however, you might try to express your ideas in writing whenever possible, rather than verbally.

I have a client with Uranus conjunct Mercury in his natal chart, whose boss always wants to think about what is being said before he responds, especially when it concerns business matters. In the early days of his employment, this not only infuriated my client, but also made him feel inadequate. He assumed his boss's slowness to respond amounted to rejection of his ideas. After a few months on the job he realized that the boss must have liked some of his suggestions because a number of them were implemented. So the feelings of inadequacy did subside, but the frustration with the slowness of response did not. Unfortunately, my client could not avoid his boss if he wanted to keep his job, and he liked his job. He did have concern about his tendency to be outspoken even if he tried to control himself, so he began to write down his suggestions, hand them or send them to his boss, stating that he would check back in a few days to discuss the matter. In this way, the ideas were out of my client's system and his boss could peruse them at his leisure. It worked and is still working.

Getting back to your mental agility, besides being quick thinking with a Uranus-Mercury aspect, you might feel that your **mind is in constant motion**. It is difficult to turn off the thinking process. Or you experience periodic surges of electrical energy in the brain, so that you have times of high mental activity, but it isn't constant. Because the mind is abnormally active, people with an aspect between Uranus and Mercury can sometimes experience **insomnia**. Finding ways to externally express your thoughts, can often relieve this condition. Telling someone what you are thinking, writing down your ideas, or even talking into a tape recorder (which is certainly a Uranus/Mercury instrument) can be successful.

If you have a Uranus-Mercury aspect in your natal chart, it will always be part of your character. But if Uranus, by progression, moves closer to exactness or becomes exact with the Mercury, the intensity of the combination will probably increase. You could become more impatient and more inconsistent. But you also could be quicker and more creative. You might find that you have to increase the number of outlets you have for communications or, just as frequently, you may have to deal with relieving your overactive mind. If you are having brilliant ideas, you need to find more ways to bring them out into the world. With more exposure, you could experience more success.

If you have no natal aspect between Mercury and Uranus, but Mercury forms an aspect with Uranus by progression, the meaning and manifestation could be similar to what has already been mentioned in regard to natal Uranus aspecting natal Mercury. As with many progressed aspects formed by the faster moving planets, it would be short-term. Since Mercury at its maximum velocity moves about 2° a day, by progression, using a 1° orb, the aspect would be in range for only a little more than a year.

You would probably experience the combination rather suddenly and it might require some personal adjustment. Even though it describes a temporary state, it could change your pattern of behavior for the period of time the aspect is in range. As stated in regard to the natal combination, you might find your **mind moving at a feverish pace**. You could find that your attention span becomes shorter and notice that you are changing your mind a great deal. If this is not part of your basic pattern, and because it occurs suddenly, you could become concerned that something is wrong with you.

If you are aware of the astrological implications, however, you can take advantage of the temporary condition rather than merely worrying about it. Instead of dwelling on the inconsistency, capitalize on your newly acquired ability to think more quickly. Make use of the creative ideas that periodically come into your mind. As to the **shortened attention span**, it could be difficult for you to follow through on long-term plans, but you could accomplish a great deal if you build on one idea after another.

There may be days when you do not feel creative and may think that you are never going to have another creative thought in your life. That, however, is not true. But you should not try to push yourself into being creative. It will undoubtedly be more constructive to use those seemingly fallow periods for implementing the innovative ideas you have already had. Trying to keep your mental skills honed can be just as satisfying as constantly producing new ideas. If you do not test your ideas, you won't know if they work, and impractical ideas have little value no matter how brilliant they sound. Besides, if you try to pressure yourself into being creative when you do not feel inspired, you will probably only make yourself more and more anxious and nervous.

Transiting Uranus activating natal Mercury can have manifestations similar to those mentioned above for the progressions, but since transits often describe circumstances in addition to your feelings and personal behavior, it is possible that others may be involved in the expression of the combination. Yes, your mind may seem to be working more quickly or erratically than usual, but you might also find that others are more outspoken with you, or those with whom you have frequent contact are behaving uncharacteristically. In such cases you might question their sanity rather than your own.

Another possibility could be that **people with whom you have had no contact for some time will suddenly appear on the scene.** In other words, **when transiting Uranus is aspecting natal Mercury, you should expect the unexpected in terms of communications**. You might activate the transit by being the initiator. If you have the urge to get in touch with someone you have not seen for a quite a while, do it. In that way you may help to shape the meaning of the transit.

This can also be an excellent time to project your original ideas into the world. Externally **expressing your creative thoughts may trigger more brilliance from within**, or be added to by those with whom you are sharing your ideas. There might be some challenge from others, but with transiting Uranus aspecting natal Mercury you should be able to hold your own. You may even find it stimulating and exciting.

Transiting Mercury aspecting natal Uranus might also indicate that you can expose the world to your creative thoughts. The main difference is the time frame. Transiting Uranus aspecting your natal Mercury would be in range of a 1° aspect for several weeks, whereas transiting Mercury aspecting natal Uranus would only be in range for about 2 days. Therefore, it is possible to wait out the aspect if you choose to do so. However, you should be aware that during those 2 days you may tend to be more outspoken than usual, or others may be tactless with you. So you have to be on guard, and it might be better to put the combination to some use. This could be a good time to test innovative ideas you have on the outer world. It might lead to sudden insights. And if your thoughts seem too strange to others, it will not last long.

Instead of trying to control what you say when you have the strong urge to be outspoken, you might instead select such times to express your individuality, or discuss your views or needs in regard to freedom. If there is something in your community you would like to change, you might choose this period to write a letter to the editor of your local newspaper stating your ideas. Because you will be inspired, you might have an impact on others and start a movement that could lead to making changes you want to make.

Uranus and Venus

If Uranus and Venus are in aspect in your natal chart, you may be **unconventional in love**. Possibly you become **bored if your love life is too serene** and the excitement of a conquest may be more gratifying than an ongoing relationship. Therefore, it is possible to go from relationship to relationship, never staying with one person for very long. You may not need a steady relationship. In fact, you could feel too restricted if you concentrate on the same significant other all the time. However, there are other ways in which the Venus-Uranus motifs can be expressed.

You might, for example, find an **unusual partner** with whom you could have a steady and satisfying relationship. Or you might share your love with someone who is spontaneous or creative, and is not at all strange. Another possibility is that you might be involved with a "normal" individual who simply allows you to freely express your affection. You could be extremely content and secure in a love relationship under either of the preceding conditions. It is usually when you begin to feel trapped that you have the urge to move on. So if you concentrate on evoking the principles of freedom, individuality, creativity and excitement in love relationships, you may stay with the same person forever, if you so choose.

There are, however, ways other than through love that you could express the combination of Venus and Uranus. Venus is the planet of sociability as well as love and affection. Therefore, you might find that you are drawn to **unusual social situations**, or you might be inconsistent in that you sometimes socialize easily and at other times want to avoid social interaction. You will probably be attracted to stimulating and exciting people but after you have absorbed what you can from these individuals, you could become bored with their company. Then they are no longer exciting and you have the urge to go out to search for more interesting companions.

If you feel that there is something wrong with you because you do not maintain long-term relationships on the social scene, while others seem to belong to the same group that they had at age 6, there are alternatives to flitting around. You could look for new ways to socialize with the same people. Or if you feel envy for those who have had friends forever, you might search for one of your childhood friends and try to rekindle old feelings. Or you could accept the fact that you prefer to change those with whom you interact socially, and that there is nothing wrong with you.

Another definition of Venus is artistry and, therefore, if it is connected with Uranus in your natal chart, you could be extremely creative. **Your art might be unusual or distinctly individualistic**. It is also possible that your creativity is inconsistent—sometimes good, and sometimes not so good. This may cause you to wonder if you really have talent. If you find the right form of art, however, you might be surprised by the amount you have.

I have a client with this combination who is a graphic artist. She majored in fine arts in college but was never quite sure of her ability as an artist. After college she became a graphic artist in order to earn a living. She continued her serious painting on the side, but still questioned her talent. Even with the graphics, she felt mediocre. She was dissatisfied with her work because she believed that she lacked the spark she felt was essential for a talented artist.

Then she found the computer (ruled by Uranus). As she became more and more involved with this piece of modern machinery, it was as though her computer lit that spark of genius within her. She now has her own graphics business which is flourishing. Her success can be attributed in part to her talents, but more importantly it is due to her confidence in her abilities which she is able to express to prospective clients. These people then become clients because they know she will do a good job for them.

Her work as an artist has become a productive channeling of her Uranus-Venus aspect and her love life has improved as well. She has been married to the same man for about 10 years. There have been ups and downs in the relationship. She had two affairs in the first few years, but as her artistry became more fulfilling to her, so did her marriage.

If you have an aspect between Uranus and Venus in your natal chart, you will feel or exhibit some or all of the qualities mentioned above. Even if the aspect is not exact at birth, the traits connected with this combination will always be familiar to you. However, if Uranus by progression is applying to natal Venus, the characteristics will become stronger as the aspect moves closer to exactness.

On the other hand, it is possible that you were not born with any aspect between these planets and, in the course of your lifetime, Venus by progression forms an aspect to your natal Uranus. In this case you will be faced with behavior and attitudes that may not be part of your basic make-up so some understanding and adjustment will be required. Since Venus moves a maximum of 1° 16 minutes a day, by progression it can move more than 90° in 90 years. If it is moving forward at its maximum velocity, it will form a major aspect with natal Uranus during that time period. Using a 1° orb, it will be in range for a little less than two years, but you may experience the combination intermittently for a much longer period of time. It may be as early as 4° prior to exactness that the feelings of restlessness begin to appear.

In matters of love and affection, you could have feelings of dissatisfaction with your present situation, whatever it might be. If you are not involved in a romance, you could have a strong urge to find one. At first, because the feelings are not constant, you could ignore them, or find ways other than actively seeking romance. You might read romance novels, or fantasize about an exciting lover. But as the aspect moves closer, and the feelings become more constant, such substitutes might be less gratifying.

Since you have the urge for romance and excitement, you could go to places that advertise such possibilities. I have a client who met her present mate during a week's vacation at Club Med. Of course, no one can guarantee this will happen for everyone. But even if it does not, it is a way to make use of the Uranus-Venus combination because you can have a pleasurable (Venus) time and be temporarily free (Uranus) of responsibilities.

If you are involved in a love relationship that has gradually become less and less fulfilling, you could select the period when progressed Venus moves into range of an aspect with natal Uranus to look for someone new. The aspect could provide the impetus to get moving. If, however, you have been relatively content in your relationship and, as progressed Venus moves closer to an aspect with natal Uranus, you suddenly are unhappy, you might not want to make a drastic move such as taking a singles' vacation — at least not at the first moment you begin to feel restless. Doing so immediately could satisfy the need for excitement and freedom, but if the dissatisfaction is temporary, you could regret your actions later on. Therefore, you might want to try other alternatives first. The reading of romance novels or fantasizing about an exciting lover, as mentioned above, are two possible ways to start.

Use your Uranian ingenuity. Consider the gamut of definitions ascribed Uranus and Venus, you could come up with a number of ways to utilize their combined drives. Reading and fantasizing may work on quieting the inner urges, but there are many external choices that can be gratifying, and even acceptable. Use the time of the aspect to develop some form of creative artistry. The two-year period during which progressed Venus is in aspect to natal Uranus should be long enough for you to master some artistic skill. You might abandon your endeavors after the aspect is passed, but it will have served its purpose. And it is possible that you will have brought forth talents that are indicated in ways other than through a Venus-Uranus connection in your horoscope, but this temporary aspect is strong enough to bring out the abilities that have been latent up to that time.

You might also apply the aspect to your social life—seeking out interaction with exciting people. You could stay in your usual social realm, but add to it. You could get involved in a group with a cause (Uranus) that would give you pleasure (Venus) or where your charm or social amenities would be appreciated. Then, when the aspect has passed or you no longer get pleasure from the group, you can leave and the rest of your life is still intact.

When transiting Uranus is aspecting natal Venus, the focus is on factors in your life that involve love and affection, artistry, diplomacy and pleasure. It could also be connected with financial matters. Uranus indicates that there is a need to change something connected with one or more of these factors, or at least that the status quo should be examined and changes considered. An inconsistency or changeability is connected with Uranus so there can be difficulty in determining the precise direction you want to take or deciding what actions are appropriate. Since this aspect can be in range for a number of weeks, and more likely about two years because of retrogradation, it can be connected with times that major issues will arise in regard to Venusian matters. This is true in terms of qualities associated with any planet in your natal chart that is being aspected by natal Uranus.

Along with the inconsistency, there are also feelings of urgency connected with transiting Uranus. You may be inconsistent, but you also have the desire to make decisions instantaneously. In dealing with this planet, especially in transit, it is best to take life one day at a time.

Because Uranus is the transiting planet, the themes connected with it need not come from within. You might have to cope with inconsistency or erratic behavior in others. But, since it is your Venus that is being aspected, this inconsistent or erratic behavior could have an impact on your love life. This still may indicate that you are supposed to at least consider, if not make, changes in that segment of your life. It is just that someone else has brought the matter to your attention.

Even though you may not have initiated the manifestation of the aspect, you do not have to feel helpless and allow someone else to control your life. You can find ways to assert your independence through your Venus. Go in search of a new love or become more independent in love. Or if those possibilities are too drastic for you, at least go out and socialize or spend some money spontaneously for pleasure. As already mentioned a number of times, when you are dealing with Venus and Uranus connected with each other, you can always use the aspect by expressing your artistic creativity. It might be a time to paint that picture you have always wanted to paint, or write that story you have been creating in your mind. Do not, however, expect to spend long periods of time on these endeavors. Be spontaneous. When the mood hits you, be creative. When the inspiration is gone, go on to other activities and return to it when you again feel creative.

When transiting Venus aspects your natal Uranus, the period of time during which it is in range is so short (about two days), that it could pass by relatively unnoticed. You can make use of it, however, by choosing these times to express your independence or individuality. With transiting Venus highlighting your natal Uranus, you will probably do it well because you will be charming or tactful. And you will undoubtedly enjoy expressing this facet of yourself. If you have been considering starting some kind of social movement, you might select a time when transiting Venus is aspecting your natal Uranus, to convince others to join your cause. The aforementioned charm and tact can certainly be an asset when you are trying to influence others. But the window of action is short. When Venus is past the aspect with Uranus, the charm connected with Uranian matters may be gone as well, unless of course you also have an aspect between the two planets natally.

To sum up the combination of Venus and Uranus, whether it is in the natal chart or occurs by progression or by transit, the principles that are ascribed to the planets should be kept in mind. You should be prepared for a range of possibilities, from **restless-**

ness, inconsistency and erratic behavior to excitement, and creativity (Uranus) in matters of love and affection, sociability, money and artistry (Venus).

A number of ways to make use of the combination have been offered, but since it is a creative combination you will certainly come up with other possibilities that will be appropriate for your situation. What is most crucial is not to be sidetracked if something negative occurs before you have begun to activate a plan of positive action. No matter what is happening in your life, think about and then act upon ways to get immediate pleasure without having to deal with dire consequences. You do not want to lose sight of the future but, if you **find small ways in the present to break with tradition in terms of love and affection, sociability, artistry and finances**, it will give you immediate gratification. And this in turn can lead to bigger and better ways to make yourself feel freer and more creative, without producing havoc in your surroundings.

Uranus and Mars

If there is an aspect between Uranus and Mars in your natal chart you are **quick to take action**. When something needs to be done, you will probably respond spontaneously without stopping to think. Therefore, you will be an asset should an emergency arise. Afterwards you might wonder how or why you did what you did. You could be considered courageous or foolhardy. You might even say to yourself that you would not do that again, but should a similar situation arise you would still probably move immediately without forethought.

Although you are quick to take action, another potential quality associated with the aspect is **impatience**. You want everything accomplished yesterday, so you not only move quickly yourself, but you expect everyone else to do so as well. If others are not as fast-moving as you, you could become angry. However, unless it is otherwise indicated in your horoscope, your anger once expressed should subside. Your sudden outburst might press others into action, and if it works you could continue the practice. If instead you experience resistance or belligerence from those you are trying to motivate, you might look for some other approach to reach these people, and find another outlet for your energy.

One way of dealing with the potential anger is to have a physical outlet that you can make use of when you feel the rage building. But even if you are not feeling anger, this is such a high energy combination you should often involve yourself in some form of exercise. If you try to sit still for too long a period of time, you can begin to fidget and feel nervous. Your physical exercise need not always be the same, nor should it be part of an organized regimen. Should you decide to lift weights every Tuesday from 9 to 10, or jog every Friday from 1 to 3, you will probably find many reasons for not doing either. **Be spontaneous. Exercise when you feel the urge** and in the manner that seems appropriate at the moment.

Your quickness and your impatience to see results might cause you to be careless. **You could be in such a hurry to accomplish a task that you might not have the patience to be thorough**. You do not want to be bothered with a myriad of details. Being able to respond spontaneously is an asset, but sometimes, when you do not take the time to work through all the details of a task, you can make mistakes. Then you have to redo what you have already done and the job takes much longer than it would have if you had used a few minutes to organize in the beginning. These words are not going to make you more patient or thorough.

There are two alternatives, however, which come to mind immediately that might help in coping with the impatience. One is to **keep your tasks short** rather than initiate an elaborate game plan. If the job is broken up into small segments, you can keep your enthusiasm high because you are seeing results, and you can also catch mistakes more quickly. The other thing you can do is to **have someone else assigned to take care of the details**. You can still come up with the ideas and initiate the plan which is the part you enjoy.

The tendency to be careless can also manifest as being somewhat **accident-prone**. Your body may move more quickly than your mind, so that you can bump into an object before your brain records its presence. One way to deal with this is to condition yourself to periodically stop and view your surroundings. Then, once you have done this, you can move on. This may sound too limiting, but if you tend to trip over your own feet, it might be worth a try.

The high energy associated with this combination has been referred to in the preceding paragraphs. It was also mentioned that scheduled physical exercise might not be followed. One reason for this is the need for freedom associated with the planet Uranus. You do not want to feel limited or be tied to the same physical activity on an ongoing basis. But there is another reason, and that is that the energy level is not constant. Your **energy might come in spurts** and you may not gauge yourself very well. When you feel the surge, you may move at a feverish pace and accomplish a great deal. But it is difficult to sustain this level of physical activity and you may not be aware of the dissipation of energy until you are exhausted.

If you experience this pattern, there are different ways in which you can handle it. One is to **keep your physical activities short-term**. Then you are not only less likely to become exhausted, but you are also more likely to finish tasks that you begin. For example, let us say that you have the urge to clean your whole house. With all your energy, you could begin by pulling all the furniture away from the walls so that you will be able to clean behind the furniture without any effort. However, by the time you move all the furniture in the house, you might be so tired that you cannot do the cleaning, nor do you have the energy to move the furniture back where it belongs. So that you have accomplished nothing except to make your home look like a war zone. Instead, you should concentrate on cleaning one room at a time. And then, even if you have run out of steam before you get to the cleaning part, you have other rooms you can retreat to which are still intact.

If you were born with an aspect between these two planets, Uranus by progression would probably always be in range of an aspect with natal Mars. If progressed Uranus is applying to your natal Mars, you would be familiar with the qualities associated with the planets. But as Uranus moves closer to Mars, there will be more intensity in terms of the combined characteristics. Should the aspect at some point become exact, the year in which this occurs could be a time when you will not only feel the drives more strongly, but you might also manifest them more overtly.

You could, for example, select that period to initiate (Mars) a revolution (Uranus). Most of us, even those with an aspect between Mars and Uranus, are not so politically connected that we would try to overthrow our government. You might, therefore, decide to champion a societally acceptable cause but, more likely, your **rebellion** would be connected with some segment of your own life. You could begin to **activate changes within yourself or in your personal environment**. Changes might occur without your assistance, so you can wait to see what happens. But with Mars in the picture you should try to make things move in the direction you want them to move. Since you always have this combination you can feel comfortable with it, and it should be relatively simple to activate it.

If you were not born with a Uranus/Mars aspect and Mars is applying to Uranus by progression, you could experience the combination at some point in your life. Since Mars moves a maximum of 48 minutes a day (representing a year by progression), it could travel far enough in a lifetime to form some aspect with natal Uranus. In this case you would have to familiarize yourself with feelings and energies that you were not accustomed to. As Mars by progression begins to form an aspect with Uranus, you might notice that your energy level picks up. Using a 1° orb, this could last about 4 years.

No one would complain about having more energy. But other possibilities might not be so pleasant. The **energy surge could be sporadic** — suddenly emerging and just as suddenly disappearing or dissipating. And this **inconsistency** might make you very nervous. Along with the energy, you will probably also feel the impatience and possibly the carelessness which can be associated with this combination. Or **you could experience sudden fits of anger, sometimes lashing out at others**. If you are basically mild-mannered, you might become concerned about yourself when you manifest this kind of behavior.

Knowing when this progressed aspect is going to occur could ease your concern about what is wrong with you. If your energy is erratic or you are uncharacteristically impatient and unduly angry, you can assure yourself that it is a temporary condition and have some idea of how long it will be part of your behavior pattern.

But you can do much more than this. You can make use of the aspect while it is there.

You can prepare by planning activities or tasks that require physical effort for those times when the energy is at its maximum and back away from such activities when the energy is not there. Since we are dealing with inconsistent levels of energy, you can figure out how much time would be needed to complete the project if you were operating at maximum speed. Then add extra time for periods when you will not be running at utmost velocity so that your deadline will not create the anxiety that can be associated with the negative side of the combination of Mars and Uranus. Even if your energy level stays high for long periods of time, your patience with a particular task might not. Therefore, your attention span as well as your physical endurance should be taken into consideration in setting deadlines. **You might consider involvement in more than one physical activity so that you can alternate them**. Then when you get bored or cannot concentrate any longer on the project at hand, you can move on to another. In that way you can accomplish more and you are less likely to make mistakes.

As for the possible anger, being physically active might help to control that. But you may still find that certain people or situations automatically arouse anger in you and, if this is not the way you usually are, it can be disconcerting. It may be that be that your temporary condition does not have long-term significance. You need only cope with it for the period of time that it is in range. If this is the case, you might choose this time to **tell people off who deserve to be put in their place**, and/or **energetically and actively initiate to improve a situation**. In that way you will be utilizing the energy in a desirable direction. You could develop the habit of **walking while you plan what you are going to say or do**. This will be easier than trying to sit still while you are making your plans.

Transiting Uranus activating natal Mars will be in range of an aspect for an even shorter period of time than progressed Mars aspecting natal Uranus. Using a 1° orb it would be in range for only a few weeks, but you will probably experience it for a matter of months, and even possibly more than a year if Uranus goes retrograde over Mars. As with the progression, it can be disconcerting if you are usually a calm person. Restlessness, nervous energy, impatience and spurts of anger are common during aspects of transiting Uranus to natal Mars. Awareness of these possibilities can make you less anxious than you might otherwise be. But as with the progressions, you can activate or make use of the more desirable qualities associated with these planets.

You can **select such times to accomplish physical tasks because your energy will be high**. One major difference between progressions and transits, however, is that the urge to take action might come upon you more suddenly with transits than with progressions. Because progressions are more slow-moving than transits, you will be exposed to the energies represented more gradually. The expression of the combination may emerge suddenly with both transits and progressions, but with progressions there seems to be more time to adjust because it is in range for a longer period of time.

Because transits are faster-moving than progressions, it is not recommended that you try to form an overall game plan for taking action. It was mentioned with the progressions that you can take a walk to dissipate the energy that you feel building up and, as you are walking, you will have time to create a plan. With transiting Uranus activating natal Mars, it is better to **act without too much thinking**. Concentrate on the moment and make the most of that time. Try not to make major decisions that will have a long-lasting impact on your life in this manner, but **you can achieve a great deal on the physical plane in a short period of time.**

Transiting Mars aspecting natal Uranus is even faster-moving than transiting Uranus aspecting natal Mars and, therefore, **spontaneity is even more strongly indicated**. The aspect will only be within a 1° orb for about four days. You could feel the nervous energy welling up when Mars is as much as 2° or 3° away from Uranus, applying to it, and it may begin to wane soon after the aspect has been exact.

Although some of the same feelings and manner of behavior can be similar with transiting Mars aspecting natal Uranus, as it is with transiting Uranus aspecting natal Mars, some distinctions can also be made. Since Mars is the planet of initiative and Uranus is the planet of change, revolution, freedom and individuality, transiting Mars aspecting natal Uranus may trigger Uranian issues in your life. You may have the urge to make changes or manifest your individuality. If you are aware when such an aspect is coming into range, you can plan in advance and use the window of time when it is within 1° to take action.

If you have the combination of Uranus and Mars in your horoscope, you have a lifetime to learn to live with it and make use of it. However, if you have it temporarily either by progression or transit, you should keep in mind that the less desirable qualities might materialize. **Be prepared for anger suddenly emerging or impatience and nervous energy which can lead to carelessness**. But don't settle for this. Your awareness of these possibilities can keep you from being overwhelmed by them should they occur. But you can **move ahead by initiating changes, energetically expressing your individuality and asserting your need for freedom**.

CHAPTER 13

URANUS IN ASPECTS TO JUPITER, SATURN, URANUS, NEPTUNE AND PLUTO

The five outer planets move so slowly that when any two are in aspect to each other, they can be in range for months or years. These aspects will appear in the horoscopes of most people born during that period. Thus the characteristics associated with their combination can be associated with a large part of a generation.

Uranus and Jupiter

The **Jupiter/Uranus cycle takes approximately 14 years** and, in that 14-year period, Jupiter moves through every sign once, and two a second time. Thus every possible aspect is formed between these two planets.

The general principles associated with Jupiter and Uranus will be combined in individuals having any aspect between them. The particular signs in which they are posited will further define their meaning and will represent attitudes and behavior common to all those born during the period when Jupiter and Uranus are in those particular signs.

When Jupiter and Uranus are in aspect, we are combining the principles of expansion and optimism (Jupiter) with freedom, sudden change and revolutionary tendencies (Uranus). People with this combination are usually **optimistic** and there may be the tendency to **take risks or to over-extend themselves**. I have a client whose horoscope indicates that he is basically a cautious person; it surprised me that he often took risks both financially and in other matters. He is a strong earth person and, therefore, should be (and is) very practical in many ways. Yet I was, at first, baffled that he takes such pleasure in speculating. Then I noticed that he was born with a Jupiter-Uranus conjunction, and I realized the importance of that combination in his chart and in his life. This may be true of other aspects connecting Jupiter and Uranus but, time and again, in the charts of those born with the **conjunction**, I have seen the **tendency to take chances** in some segments of their lives.

When we add the sign(s) in which they are placed to the interpretation, we get more information that can be applied to a great number of people. Since we reflect the quality of the time in which we were born, we might look at what was occurring in the world during a period when Jupiter and Uranus were aspecting each other and apply their meaning to those who were born during that period.

An excellent example of this is the **conjunction of 1927-28**. The optimistic attitude was certainly evident at that time. The slogan of the Republican party for the Presidential election of 1927 was, "A chicken in every pot, a car in every garage." The populace was evidently in the right frame of mind for the slogan and elected the Republican candidate.

During this period, **enthusiasm was high** and people experienced feelings of well-being. There was no worry about tomorrow. The optimism led to **extravagant spending,** and the element of risk-taking which was evident with people **buying stock on margin**. The conjunction of the 1927-28 period started in Pisces and then moved on to Aries. Either of these signs could compound the optimistic and "devil-may-care" attitude. With Pisces emphasized, the element of fantasy would certainly help to foster the optimism of Jupiter. Few would take time to analyze if it were feasible for everyone to have a chicken in every pot or a car in every garage. It sounds good and there seems no need to make sense out of it. Then with Aries, the element of risk-taking is reiterated. With this sign, there isn't the patience to think through all the consequences and wait for tomorrow to take action.

The main difference between the attitude and behavior of those born with Jupiter and Uranus conjunct in Pisces and those with it in Aries is that those with it in Pisces would tend to speculate or dream about the way things are or should be, and those with it in Aries might try to make them happen. Therefore, **optimism would be more evident with the Pisces people and risk-taking with the Aries**.

If we look at the signs of the **conjunction in 1968-69** in connection with world events we see different facets of Jupiter and Uranus. A major event of 1969 was "**Woodstock.**" It may have seemed to be merely a giant rock concert if viewed superficially, but it was really a mini-revolution—the start of a counterculture movement. When people who attended the event are interviewed about the experience, many of them say it had such an impact on them that they began to grow personally (Jupiter) and important changes (Uranus) took place in their lives. And they also discuss the camaraderie and sharing they had with others who attended. It should not be surprising that the conjunction was in Libra in 1969—the sign of partnership and artistry. Those born in 1969 with the **conjunction in Libra**, therefore, will often **involve others when they contemplate risk-taking or making dramatic changes in their lives**.

It should be evident from the preceding examples that the signs in which the planets are placed provide added insights into the meaning of the combination. The conjunction, of course, is the strongest aspect and involves only one sign so that it is a simple matter to find world events that are appropriate to the planets and the sign. The method should be applicable to other aspects (particularly the hard ones which are considered aspects of manifestation), but in these cases you would add to the complexity with two signs instead of one. This step is left to the reader.

Looking at progressions between Jupiter and Uranus (and for that matter those between Uranus and Saturn, Neptune or Pluto) would add little information. Jupiter moves a maximum of 12 minutes a day and, therefore, it would cover no more than 18 degrees in 90 days (representing 90 years by progression). Thus, if you were not born with an aspect between Jupiter and Uranus and progressed Jupiter moved into aspect with natal Uranus, it would move so slowly that you would have time to adjust to it and it would have much the same meaning as the natal aspect for the time it was in range. If you were born with an aspect between Jupiter and Uranus and it becomes closer to exact by progression, you could find yourself becoming more daring, more revolutionary, more willing to openly express your individuality. If the aspect is separating, but does not move completely out of range, you might become a little more cautious and less apt to take chances as you become older, but probably no one but you would notice!

When we discuss transits of either Uranus or Jupiter, there are certain similarities these two planets have. Both are connected with the concept of **exploration**. So whether transiting Uranus is aspecting natal Jupiter or transiting Jupiter is aspecting natal Uranus, there could be an urge to take risks—**to try new and different things**. This combination in either direction could also coincide with a **sudden windfall**. You could wait and see if this works for you, or you might select a time when transiting Uranus is about to aspect natal Jupiter or transiting Jupiter is aspecting natal Uranus, and do something that might bring you a windfall. For example, you might buy a lottery ticket each week while the aspect is in range. You should not, however, get carried away with this idea. It would be foolish to take out a second mortgage on the house in order to buy a hundred lottery tickets each day or week. No one can guarantee that you will definitely get a windfall, and there are more productive ways to use the combination. But an occasional small risk can satisfy the urge for excitement and, who knows, it might pay off.

Although there could be the impulse to "throw all caution to the wind," it would be safest and possibly most rewarding, if you explore one step at a time. **The combination can be overwhelming if you move too far, too fast**.

There are also certain distinctions that can be made between transiting Jupiter aspecting natal Uranus and transiting Uranus aspecting natal Jupiter. First of all, the length of time in which they are in range is much different. Jupiter, when it is moving forward at its maximum velocity, is in orb of a 1° aspect with Uranus for about ten days. Transiting Uranus is in aspect for months. Aside from the length of time in which the aspect is in range, there are certain qualities associated with each of these planets that can help to differentiate them in transit.

There is **more restlessness and urgency connected with transiting Uranus** than with transiting Jupiter. So if transiting Uranus is aspecting natal Jupiter, you would **not feel consistently focused**. You want to expand and develop, but you would not have the patience to form a game plan before taking action. Therefore, you could find yourself flitting from project to project or trying to expand in a number of directions at the same time and accomplishing very little. But, on the other end of the gamut, you could find that every once in a while you experience a **spark of genius** in regard to your expansion, and you will have a **sudden success**.

Your best course of action would seem to be to sit back and wait for that "spark of genius!" But with transiting Uranus activating your chart, it is hard to sit and wait. Besides, if you are not exploring, you might not have the stimulus to ignite the spark. Remind yourself of this when you feel that your behavior is erratic or unproductive. You also have other alternatives when transiting Uranus is aspecting natal Jupiter. Besides representing development and expansion, Jupiter is also connected with fun. Therefore, you could use some of the time during this aspect to create some **spontaneous pleasure**. This might help to **renew your vitality** and start the creative inspiration flowing again.

When transiting Jupiter is aspecting natal Uranus, there could be the desire to **express your individuality or revolutionary tendencies in a big way**. The Uranus facet of your character will not easily stay in the shadows during this time. You could make a statement about your need for freedom by finding ways to express your independence. You could **embark on a new path** that might broaden your horizon. Keep the idea of expansion in mind as you attempt new ventures.

Whatever else is occurring astrologically, when you have Uranus and Jupiter connected either in your natal chart or by transit, **excitement and development** will be important to you.

The main difference is that if the combination is in your natal chart, you will experience the feelings throughout your lifetime, although possibly sporadically. If it occurs by transit, the condition will be temporary.

What happens is that if your life is too sedate, you could feel bored, dissatisfied and want to do something about it. If you live with this combination in your horoscope, you can learn to recognize the symptoms and use these periods to **start new ventures or create some harmless excitement for yourself**. This combination has nothing to do with will power or common sense, so you might sometimes find yourself in situations in which you are overwhelmed because you have gone too far without evaluating, or even thinking about it. If this happens a few times, you will probably learn to recognize the feelings at an earlier stage. Then you can have plans ready in advance which you can activate when the feelings first emerge. Once you consciously begin to make use of the energies, you could find that this combination is an asset in helping you to expand and develop as an individual.

If the aspect that occurs by transit (either transiting Uranus to natal Jupiter, or transiting Jupiter to natal Uranus) is not present in your natal chart, the urge for excitement and freedom may be new to you. You may be hesitant to take steps to make any changes because you are uncomfortable with these new feelings. If you cannot (or do not) make changes, you could become dissatisfied with yourself or your situation. It may be that you believe that others are interfering with your development or you may think that you cannot control your own behavior. But now that you are aware of these possibilities, you know that you can use these times to make giant strides forward.

Uranus and Saturn

Saturn has a cycle of 29.46 years and within this time frame will form every possible aspect with Uranus. The **conjunction occurs approximately every 45 years**. In this century it took place in **1941-43 in Taurus and Gemini, and in 1987-89 in Sagittarius and Capricorn**. The principles involved are responsibility (Saturn) and freedom (Uranus). Those who have this combination in their natal charts undoubtedly feel pulled between doing their duty and needing to feel unencumbered by responsibility. Both qualities represented by these planets should be accommodated, but one may be stronger than the other because of the signs in which these planets are posited.

If you keep in mind the definitions of the planets and the signs, it is a relatively simple matter to determine which planet is likely to be stronger in each sign. When the conjunction occurred in Taurus, Saturn would feel more comfortable because Taurus is the fixed earth sign, and Saturn rather than Uranus would prefer the slow, steady pace associated with Taurus. It is easier to do what you are supposed to do in such an environment. Whereas in flighty Gemini, Uranus would tend to be stronger because it is more effortless to make changes and express individuality than it is to stick to the straight and narrow.

The same comparisons can be drawn with the signs of the 1987-89 conjunction. The conjunction in Sagittarius is a more comfortable placement for Uranus than for Saturn because Sagittarius is associated with expansion, exploration and freedom. These are key words with which Uranus can feel at home. Whereas the conjunction in Capricorn, the sign Saturn rules, would, of course, be more comfortable for Saturn than for Uranus.

Armed with the information of these planets in their respective placement in signs, you can be aware of which planet will most likely be easily expressed and which will require effort. You can continue to utilize the stronger planet as you usually do but at the same time you can take small, scarcely noticeable steps toward expressing the seemingly more ineffective planet.

If Saturn is the weaker planet, undertake a commitment from which you can see the results in a short period of time and doesn't interfere too much with your freedom. For example, your office might be a mess and you may have been postponing putting it in order because it is such an enormous task. You might set a fifteen minute period and see how much you can accomplish during that time. Who knows, you might feel so good about what you achieve that you decide to take another fifteen minutes and, before you know it, you have cleaned the whole room. Once you begin the pattern, it might grow and eventually you will be balancing the Saturn and the Uranus energies in your life. Even if you do not spend more than fifteen minutes straightening up, you could find that you have cleared enough space to accomplish more work. Or you can go back to doing your own thing without feeling guilty.

If the definitions associated with Uranus seem to be unexpressed in your life, try to find ways to be creative. Force yourself to do something you consider a little daring. You do not have to go bungee jumping. If Uranian energies have been subdued, any small disruption in your routine could be enough. Take a couple of hours and go off by yourself or attend a creative workshop. You are not looking for patterns to establish. Try something that is freedom-oriented once. If you like it, you may repeat it sometime. If you don't like it, you don't have to do it again. By pushing yourself to do a small, spontaneous act occasionally, you could awaken that spark of creativity within you, or at least alleviate the feeling that all you do is your duty.

No matter in which signs they appear, however, whenever Saturn and Uranus are in aspect to each other, freedom and responsibility are joined together. It is true that they may interfere with each other, but once you understand that both have to be expressed, you can begin consciously to bring out the qualities of both. If you have an aspect between Saturn and Uranus, you may have a strong urge to free yourself of responsibilities or get rid of commitments almost immediately after you have accepted them. But the principles of both planets are part of your total character and each must be utilized in order for you to feel complete. If you never accepted responsibilities you could feel free, but you might not believe that you are worthwhile. If you always do your duty and do not allow time for being your own person, you could think that you were dutiful but your creativity could be stifled.

You might alternate the two by taking on a commitment (Saturn) and following through to completion. Then you can reward yourself with time off to feel free and express your individuality and creativity (Uranus). This will work well if you keep your commitments short-term. So you should set deadlines for yourself.

By setting deadlines for yourself you can avoid the feeling that you are trapped forever as soon as you make a commitment. This does not mean that once the date of the deadline is reached you must drop the commitment. What you should do is to evaluate your position. You might decide you have done enough, give yourself some time off, and then move on to another commitment. Or you might decide to extend your commitment.

I have a client who was born during the 1941-43 period and has the conjunction in his seventh house. He told me that after I explained the possible meaning of the aspect he understood his attitude and felt better about himself. He also said that he has begun to handle all his commitments in the manner just described —even his marriage (the seventh house placement). He informed me that once a year he evaluates his relationship with his wife and, so far (he has done this twice), he has decided the relationship is good enough to reaffirm the commitment. He admitted that he has not informed his wife that he is doing this. However, from his perspective, he feels that this evaluation has helped him to appreciate the marriage more than he did before and strengthened his commitment to his spouse.

You need not become obsessed with balancing freedom and responsibility. There are ways in which the energies of Saturn and Uranus can be expressed without using either of these definitions. You might use Saturn to establish rules and when particular situations arise, determine which rules must be followed (Saturn) and which can be broken (Uranus). Another possibility I remember reading is that this combination can be associated with "earthing" (Saturn) of brilliant ideas (Uranus).[3] So if you feel heavily weighted down by your responsibilities or unsettled by your lack of commitment, remember this ability that you have. **Take some of your creative thoughts and make them materialize in a physical way**.

Or you might **get involved in a cause (Uranus) and make things happen in regard to it (Saturn)**. If you keep in mind practical goals and a definite course of procedure, you can be successful. Then if you join forces with others who operate in this manner, as a group you can have a profound impact.

By progression, Saturn moves a maximum of eight minutes a day (representing a year by progression), so in 90 years it would cover about 12° traveling at maximum velocity. So if you did not

---

3   Baigent, Michael, Nicholas Campion, Charles Harvey, *Mundane Astrology*, Wellingborough, England: The Aquarian Press, 1984, p. 181.

have an aspect between Saturn and Uranus in your horoscope and Saturn is the planet approaching Uranus, you could have an aspect formed during your lifetime. Its description would be the same as given above. It would just start a little later in life.

In terms of transits, however, there is a difference between transiting Saturn aspecting natal Uranus, and transiting Uranus aspecting natal Saturn. When Saturn activates planets in transit, you are informed that it is time to **clarify, correct or manifest** something that can be identified with the natal planet it is aspecting. Or, the part of your make-up associated with the natal planet could be limited or frustrated. And you could experience all of the above during the time of the aspect.

Since Uranus is connected with your individuality, freedom, rebellious tendencies and creativity, as transiting Saturn aspects it, you should be working to understand the meaning of these qualities within yourself and find positive ways of expressing them. When Saturn is the transiting planet, as it moves into orb of an aspect, you frequently become aware of your limitations in regard to the natal planet that is being aspected, especially if you have been ignoring that facet of your life.

When transiting Saturn aspects natal Uranus you might find that someone makes demands on you that interfere with your freedom. You might not have been expressing the freedom principle in your life, but now that you are being pressured to take on a responsibility, you realize you want to make use of it. If you are feeling free and creative, when transiting Saturn appears on the scene, you probably reap rewards. So as Saturn begins to come into orb of an aspect with your natal Uranus, examine and clarify your needs in terms of Uranian issues. If, no matter what you have tried, you become aware of the aspect because of limitations placed upon you, do not consider the situation utterly hopeless. You can always work out clearly defined steps to take so that you can begin to redirect the inner drives.

When transiting Saturn is aspecting natal Uranus, you might have an experience whereby you feel that another person finds you boring, or at least unexciting. This can be depressing and you may simply wallow in self-pity (one possible way of using Saturn). Or you might decide to show that person that you are exciting and display your individuality. You could dye your hair purple or wear an outrageous outfit. This should certainly make others notice you, but you have to decide if this is the way you want to be seen. With Saturn, you might want to consider alternatives before you take action. There will be the urge to make sweeping changes immediately, but you will be less likely to make mistakes if you plan first, try small changes and evaluate. You could, for instance, wear your hair in a new style and see if anyone notices. If so, decide if this is the response you want. If it is, it may be enough to make you feel that your individualism is respected. If not, it is much easier to try a new hairstyle than to remove purple dye or alter more drastic actions! In other words, with Saturn you should appraise what is occurring as you go along and not wait until you feel totally restricted.

Although you may believe that you are not being respected as an individual during an aspect from transiting Saturn to natal Uranus, the concept of freedom will probably become an issue as well. You might become aware of your need for independence when a responsibility is put upon you. Therefore, you can wait until it occurs, or being aware that it is a possibility, you might begin to investigate, before Saturn moves into range, what responsibilities you can take on with the least loss of your independence. Then, when the time arrives you volunteer to take on the commitment you have chosen. It might be a good idea to **keep the commitment brief**. In this way, if others ask you to do something for them, you have a reason to say no without feeling guilty. If you want to be of service to them, you can take on the responsibility sooner without feeling that you have failed to follow through. You will also be **less likely to feel trapped if the end of your responsibility is in sight**. And, it is always much easier to extend your commitments than to end them.

But there is yet another way to bring out the qualities of transiting Saturn aspecting natal Uranus and that is by cultivating creative talents that could distinguish you from others. During this period you could have a strong desire to try your ideas in the real world. Before the aspect moved into range, you may have been satisfied with just creating, but when Saturn appears you feel you have to **make use of what you have conjured up**.

In spite of that, however, you could worry (a quite common occurrence with Saturn) about the results. Your concerns could be about the practicality of your ideas. What if they are impractical? If you cannot make use of them, you may question your creative ability. You might consider it safer not to take any action rather than expose your ideas to the possibility of criticism or failure. And should you take this route, it will be highly likely that you will feel uncreative and experience all those negative Saturn definitions.

Even though it may take effort, you will ultimately find it more gratifying to expose your ideas to the outer world than to be totally inactive. You do not have to move fast. You need only move. **Take small steps and evaluate your progress as you do**.

I have a client who had invented a gadget and selected the period when transiting Saturn was opposing her natal Uranus to make the prototype and to seek financial backers. The task was tedious. She talked to a number of manufacturers, weighing price against quality. She also did a great deal of research to find suitable backers. Her invention was original (Uranus), but it took Saturnian effort to make it materialize. She has also felt that the whole process was very slow. The product has been on the market for two years now. It was not an immediate success, but it is steadily becoming more and more profitable. And the best part is that **what you earn under Saturn stays with you for a long time**.

The specifics of this story may not be applicable to you because not too many people have inventions in their minds that are waiting to be born. The principle, however, is relevant. The example illustrates that under Saturn, hard work can produce results. It should be added that, since Uranus is the planet being activated, you will not have a great deal of patience to wait for success. What worked well for my client were **periodic diversions**. At times we discussed creating some excitement in her life. At times it happened spontaneously. In other words, she was able to stay on track by taking short side trips on the way.

Another way to use the period when transiting Saturn is aspecting your natal Uranus is to **work on methodically developing creativity within you**. You may not feel creative, but Uranus is in your horoscope and, therefore, you have to have some form of creativity. Since Saturn represents, among other things, the teacher, you could use this time frame to develop your talents through workshops or classes. As with anything Uranian you do not want to commit yourself for an extended period of time, but try anything that appeals to you with the idea that you are just going to explore. You may be surprised at what you produce and what you find out about yourself. If you do not try to bring out the Uranian part of you in some way, you will probably feel limited and frustrated.

Transiting Uranus aspecting natal Saturn, however, gives a different message from transiting Saturn aspecting natal Uranus. It suggests that it is **time to change your life structure**. Sometimes your material world changes because of someone else. You may believe that your life is going well and another person suddenly does or says something that changes your environment. Perhaps your spouse moves out of the house. The structure of your home has changed without active participation on your part, but you have to deal with this change. Or, you may have a feeling that you should make changes, but you decide that you will try to maintain the status quo, possibly because you are afraid of the unknown. Then someone gives you an ultimatum—change, or else! It may be that the no one is forcing you to make changes, but you feel that change is inevitable. You might want to keep your situation as is, but it does not seem to be an alternative. It might be wise for you to make use of the aspect before someone else does. The restless urge associated with Uranus will certainly be evident. You will want to do something. You might start with minor changes that do not threaten to totally disrupt your life structure.

As you see transiting Uranus beginning to form an aspect with natal Saturn, you begin to think about meaningful changes you would like to make in your life structure, and start to implement them during the time of the aspect. It will be most comfortable to **make small changes and wait for the results**. If they are favorable, you can go a step further. If not, you can try something else before you go too far on any path.

This approach can be applied to commitments and responsibilities as well as to your physical world. When transiting Uranus activates your natal Saturn, responsibilities can begin to weigh heavily, even if you have not previously felt this way. The main message of the aspect is to **change your commitments**. This can mean merely a change in attitude or a shift in emphasis on the one hand, to totally freeing yourself from some responsibility on the other. With this aspect, you may initially feel that you must take drastic action immediately, whereas all that is really needed is a slight shift of direction. Therefore, as with all Uranus aspects, there should be **exploration and tentative changes made before irrevocable action is taken**.

You might **take a leave of absence** from a commitment before you totally let go of it. When you are constantly facing the responsibility, the natural Uranian response is to think only about freeing yourself. When you are away from it, you may become more objective about it. There is an objective quality associated with the planet Uranus that can be utilized when you are not directly involved. If you can detach yourself from a situation, you are better able to view it from different perspectives. In so doing, you might discover that a break from the commitment is all that you needed. Then you can return to it with a clear mind and an assurance that you want to continue with it. Or as you evaluate your commitment from a neutral position, you could find that you are better off without it. Then you can make your leave of absence permanent, knowing that you have made your decision rationally and not in the heat of the moment.

Uranus to Uranus

We all have only one Uranus in our natal charts, but both progressed Uranus and transiting Uranus can form aspects with your natal Uranus. If Uranus changes direction by progression in your lifetime, it is possible that it could form a conjunction with your natal Uranus. If this occurs, you will find that Uranian issues become prominent as the aspect moves to exactness. And the year during which the aspect is exact will probably be a year you will remember. You could experience either internal or external changes and you may find that matters involving freedom, rebellion or individuality are prevalent.

If you were born with a retrograde Uranus, your revolutionary tendencies might have been subdued or slow to develop, but both the year during which it turned direct and the year the conjunction becomes exact will be times of particular importance in regard to externally expressing Uranian motifs. These **could be periods when you openly rebel, or you are recognized as being independent or unusual**.

If you were born with a direct Uranus, the year in which it turned retrograde and the year of the exact conjunction could be times when you internalize the Uranian message. These may be years in which you discover that making changes within yourself is more important than dramatically changing your external environment, and that inner freedom is more crucial than making a statement about freedom to the outer world.

Because it moves so slowly by progression, it will probably not manifest suddenly. It may become more evident in your life during the years it changes direction or forms an exact conjunction with your natal Uranus (if either occurs) but it will happen gradually. Since Uranus moves a maximum of 4 minutes a day (representing a year by progression), it can be in orb of a 1° aspect with its own natal position for years.

Uranus in transit is more intermittent than it is by progression. **Since the Uranus cycle takes about 84 years, it forms a hard aspect (square, opposition or conjunction) with natal Uranus about every 21 years**. So the restlessness, urgency and/or other feelings and qualities associated with the planet can be strongly evident during those periods. And issues of freedom and individuality will often be in the foreground. You will probably notice that such feelings and issues appear suddenly for short periods of time long before transiting Uranus moves to within a 1° orb of natal Uranus.

The first square occurs between the ages of 18 and 22. As psychologists discovered in studies of the adult cycle of life, most people, at some time during those ages, will be grappling with a need to **express their independence** or a desire to free themselves from something in their lives. The specific circumstances of life may be different for each individual but the general principles are similar.

In our society, this seems to be a time when we are expected to express our independence. People around the age of 18 usually graduate from high school and either go to college or work. Some individuals leave their familial home to go off to school or move into their own apartment. Whether your parents are supporting you (as could be the case if you are attending college) or not, you would still find that you are freer because you do not have to give an accounting of your actions to your parents on a daily basis. If you go to work but do not leave home, you become more independent anyway because you are earning your own money and, therefore, probably do not have to ask your parents for financial assistance or even consult with them before spending your money.

Aside from the need for more independence during this first Uranus square Uranus, you may also have a strong **urge to explore**. The restless quality that frequently occurs during Uranus aspects is potentially doubled when transiting Uranus aspects its own natal position. There is a need for excitement and often a burning curiosity to investigate uncharted areas.

This aspect occurs at about the same time as **transiting Saturn squares natal Saturn (ages 20-23)**. So very different principles might be occurring simultaneously and you could be faced with the **freedom-responsibility dilemma**. If the Uranus square Uranus appears first it could be considered a prelude to the Saturn aspect. You would have time to satisfy your urge for freedom and exploration before it was time to take on the responsibility and commitment that are associated with Saturn. However, if the aspects occur at the same time you should find ways to express both.

It is important to accentuate each of these seemingly conflicting sets of needs. If you were only to explore and search for excitement, you might be satisfying your desire for freedom. But without any responsibilities or commitments you might begin to wonder about your self-worth. On the other hand, if you were only to work on establishing yourself in the material world without expression of the exploratory element of your personality, it might be difficult to ignite the spark of creativity that is part of every horoscope. You might begin to feel bored and boring. And eventually you could start to wonder what you might have missed. Even if you attain material success, you could feel that something is lacking in your life.

You will be aware not only of the imbalance if it should occur, but you will know which aspect is dominating because of your symptoms. Chances are you will feel restless and/or be erratic and inconsistent in your attitude or behavior if you are not making use of the Uranus. If you are feeling depressed, repressed or restricted, it is Saturn that is being ignored. It is natural during Uranus square Uranus and Saturn square Saturn to have a little bit of all the feelings mentioned above. It is only when the negative qualities associated with the planets persist for extended periods of time that you are probably not making use of the energies assigned to that particular planet.

You should not, however, wait until you become extremely nervous (Uranus) or depressed (Saturn) to do something about it. The longer you wait, the harder it is to take action. As soon as a symptom appears, find ways to bring out the positive side of the planet. For example, let us say that you get a job and are working hard to establish yourself in the material world. Perhaps you are working so hard that you have no time to do anything else. This is when you will probably begin to feel restless. So you work harder and, the harder you work, the more restless you feel. You could become so nervous that your work starts to suffer. Then you could defeat the Saturnian goals you are trying to achieve.

Instead of ignoring those urges, find ways to accommodate them. Allow time for a spontaneous excursion. You can tell yourself that the purpose is to refuel so that you can accomplish more. And it is true that if you take a break from your work, you are less likely to become exhausted. Therefore, in the long run you will be more efficient, and efficiency can lead to greater success.

If, on the other hand, you are avoiding commitments and only expressing your independence or satisfying the desire for excitement, you could feel worthless, become depressed. You can combat these feelings by taking on responsibilities and working toward goals. When you see tangible results from your efforts, you will develop more self-esteem. Intellectually, it is easy to accept this idea, but to do something about it is another matter. If you have established a pattern of noncommitment, your emotions can take over even when you just hear the words responsibility and commitment. You could become anxious and start thinking of ways to avoid consenting to anything that will be confining. Or if, in a weak moment, you have accepted a commitment, you will probably be spending more time trying to extricate yourself than on fulfilling the responsibility.

If the preceding scenario is all too familiar, you can begin to turn it around with a minimum of effort. Instead of worrying about being restricted forever, **select one small task each day for which you can see tangible results**. For example, you might start by deciding that you will wash your dishes each morning at 8 AM. After you have done this, you can note how nice the sink looks without a pile of dirty dishes, and you can see that you have done something that is worthwhile. Once you have performed this chore and evaluated the results, you can then go on with your life as usual. This may be a small start, but you can build on it. Through practice you will probably become more efficient at the initial task you selected. Then you can add a second chore and a third, which ultimately will take no more time than the first task took when you started to do it.

When **transiting Uranus opposes its natal position between the ages of 39 and 43**, there is no square, opposition or conjunction of transiting Saturn to natal Saturn. Instead there is a **square of transiting Neptune to natal Neptune occurring during this period (ages 40-43)**. So you again experience the Uranian restlessness and desire for freedom, but this may be coupled with some confusion, or feelings of dissatisfaction and disappointment with your position in life that can be associated with the Neptune square Neptune.

In our society, psychologists have called the **age of 40 as the onset of the mid-life transition (or mid-life crisis)**. By this age you have supposedly established yourself in the world and may have become comfortable and successful in the roles you perform. But somehow around the age of 40, no matter how successful you are, you could begin to experience the **restlessness** that astrologers connect with Uranus, and dissatisfaction that astrologers associate with Neptune. If all is going well in your life, you may wonder why these feelings are emerging. There may seem to be no external reasons for them, yet they are there.

One of the advantages of knowing astrology is that you can quickly understand, at least symbolically, why something is happening. As astrologers we should be able to accept the meanings ascribed to Uranus opposition Uranus and Neptune to square Neptune. We need not dwell on the whys and wherefores which can be time-consuming and unproductive. Rather we can concentrate on bringing out the qualities that will be beneficial and gratifying.

Those not familiar with astrology may need more tangible reasons. One of the "reasons" that can be associated with Uranus opposition Uranus is that around the age of 40 you begin to realize that your life could be half over and you want to kick over the traces and do something exciting. Another is that possibly your children (if you have any) are grown and they leave home. Then you are freer than you have been for years (whether you want to be or not). You may welcome it with eager anticipation, or you might feel emptiness at your loss, or dread in regard to the void their leaving has created.

You are supposed to be **making changes** when Uranus is in the picture and if you try to maintain the status quo, others may force you to adjust. As mentioned above, your children's departure from the home has to create a different atmosphere in that part of your life. Another possibility, if you are married or involved in a serious relationship, is that your spouse or significant other makes use of your aspect (or if this person is around your age, he or she might also have Uranus opposite Uranus) and wants to change the relationship or even be free of it.

Perhaps, while you are trying to keep the relationship operating in the usual manner, your spouse has an affair. If you find out about it—and ordinarily you will suddenly discover it when transiting Uranus is opposite natal Uranus—you will have to deal with it in some way. If you choose to try to ignore it, your spouse may make decisions for you. Then you undoubtedly will feel that your individuality, your need to be in charge of your own life is being interfered with, and the harder you try to hold onto the relationship, the more dissatisfied you become with your situation.

Even if you are aware that you, personally, want to make changes and no one else is pressuring you to do so, you might not select the most satisfying way of expressing your urge for freedom. You could have a strong desire to break away from your established patterns and do the unusual. Let us say that you have been married for years and considered yourself relatively happy. And in spite of this you become involved in an affair. There was no planning or forethought. It just happened. Then you have to deal with the aftermath which is not always pleasant. As one client told me, "I was looking for freedom and all I got was another obligation!"

Being forewarned can be forearmed. If you are aware of how you might feel and what can occur during the mid-life transition, you will undoubtedly deal with it more effectively. Instead of worrying about what is happening, you can develop a beneficial mind set. Accept the fact that this is a time during which you should make changes. Both astrology and psychology can give you this message. But with astrology you will not only know that this is a time when you should change, you can also see where the changes should occur along with some of the choices you have to implement the changes.

For example, if your children have left home and you are feeling the emptiness or dread mentioned above, you can choose to do nothing but feel bad. Or you might force yourself to try new experiences, explore the world around you. After a while you may discover that you not only are having a good time, but you are stretching your boundaries and creating a more exciting world for yourself.

If your marriage or serious relationship is boring to you or seems to be boring to your partner, **create a little excitement** within that relationship before you complicate your life with another person. Be spontaneous, add some romance.

The transiting Neptune squaring natal Neptune occurring during this period can be used along with Uranus to improve the relationship. Fairy tales and illusion fall under the realm of Neptune and, therefore, might assist you in instilling a little romance in your life. It could take some effort, but it is better than just being disgruntled.

If you further examine the purpose and possible meanings of Neptune you can determine fruitful courses of action. Since Neptune is associated with spirituality, it is understandable that material success is no longer gratifying. Material rewards cannot compensate for the intangible kinds of success. A pile of money cannot replace immortality. During Neptune square Neptune you may get in touch with your own vulnerability. Therefore, you could want to produce something that will live after you. The mundane world seems less important.

When you combine the meanings of Uranus and Neptune, it is clear why you frequently hear stories of individuals who, in their late thirties and early forties, suddenly (Uranus) become extremely religious (Neptune). Or those who have been societally oriented leave lucrative careers to search for their gurus.

It is natural to **want to create a new reality** during this period. Psychologists have discovered that the mid-life transition is often associated with many kinds of change — sometimes drastic, sometimes subtle. There might be conspicuous changes such as a divorce, a new job, a change of residence. But even if there is not a visible shift in the life, "If we look more closely...we discover seemingly minor changes...[have made] considerable difference. A man may still be married to the same woman, but the character of his familial relationships has changed appreciably for better or worse. Or the nature of his work necessarily changes in certain crucial respects during the Mid-life Transition."[1]

Common sense is not associated with either Uranus or Neptune, so it is best to **make small changes first**. You can then see how you feel about them before you move further. If you move too far too quickly, you could put yourself into an untenable position. For example, let us say that you have a strong desire to get in touch with the spiritual side of your nature. You quit your job and go off to join a religious movement. After a few weeks or months, you decide this was a poor choice. What can you do about it? Whatever you choose to do will take time and effort. If, however, you had taken time to investigate the religion on week-ends or during vacation time, you could have come to the same conclusion without having disrupted your entire life.

In other words, it is possible to make small or tentative changes (Uranus) and search for spirituality on a part-time basis, or temporarily dissolve (both spirituality and dissolution are connected with Neptune) extraneous factors in your life, before you make more irrevocable changes.

A safe and productive way to use these aspects is to **do something artistic**. Uranus represents creativity and Neptune is considered the higher echelon of Venus, so artistic endeavors can be successful. Since you can create your own reality during Neptune aspects, you may find that you can express talents you never knew you had. It is possible to activate previously latent

1. *Seasons of a Man's Life*, p. 61

abilities that you might never have made use of had you not experienced the Uranus opposition Uranus and Neptune square Neptune. One client going through the mid-life transition decided to learn to play the guitar to relieve the confusion and nervous energy. It turned out, according to her instructor, that she was very talented. She learned quickly, played well and even began to do some composing. This not only made her happier, but it also improved her self-esteem.

There are, of course, many ways to make use of the themes associated with Uranus and Neptune, and it is impossible to mention every one. But, if you keep the principles associated with these planets in your mind while you are experiencing the aspects, you may find different ways of expressing them without disorienting yourself.

The next hard aspect of transiting Uranus to its natal position is the **second square of the cycle and occurs around the ages of 61-62**. Freedom can again be an issue at that time. If you have been a member of the work force, you might be thinking about retiring from your job. This idea is supported in our society by the fact that you can start collecting Social Security at age 62. Often companies offer incentives for retirement before the age of 65. So you may be thinking about the possibility of **newfound freedom**, as with the Uranus opposition Uranus of the mid-life transition, with either eager anticipation or dread of the empty hours that will have to be filled.

Often clients who have worked all of their lives will come in for readings at the time of this Uranus square Uranus, or just before it, and have concerns about retirement even if they believe retirement is a few years off. They will say such things as "I don't know what I'll do when I don't have the office to go to" or, "I like the idea of being free, but will I have enough money to do the exciting things I would like to do after I leave the job?"

Once the issues connected with retirement begin to appear, you should do something about it other than worry. And "doing" something will help alleviate the feelings of anxiety. If you are worried about filling the empty hours after retirement, explore free-time activities that appeal to you before you retire. You might enjoy your new interests so much that the transition will be much easier than expected. Or, if you think you won't have enough money for the exciting things you hoped for after retirement, why wait? Don't postpone those wonderful experiences. If you cannot afford them later on, you will at least have memories. But you might be surprised at how much you can do after retirement on a smaller income. Often prices are reduced for senior citizens and there are many programs designed especially for them that are extremely interesting and inexpensive.

If you live long enough to experience the **Uranus return, which occurs around the age of 84, independence can become an issue** again. I have a few examples of people who lived through their Uranus returns. In some cases health was an issue during the time of the return, and in others, the native's children were concerned with whether their parent was capable of taking care of him/herself. In each case, the approach that worked best was to **find small ways to express independence** which led to more changes and made the individual feel freer.

One client was going through the Uranus return as she was recovering from a serious illness. She had always been a very independent person and resented the fact that she had to rely on her husband to do so much for her during the period of the illness. As she began to recover, although she was weak, she would consciously select **something new to do for herself each day**. By pushing herself in this manner, she was back to her old self in a very short time.

Uranus and Neptune

The Uranus/Neptune conjunction occurs in the sky every 172 years, and looking at what has been occurring in the world during the conjunction of the 1990s, and noting the impact of the combination in transit on clients, provides a wealth of information. Through observation, you can not only better understand how these planets may manifest together in an individual's life, but also you can form ideas of ways to make use of them both natally and in transit.

In the preceding section, Uranus and Neptune were discussed in the mid-life transition period. But we were talking about using them side by side (Uranus opposition Uranus and Neptune square

Neptune), not in combination. When they are aspecting each other in a natal chart it is somewhat different. We are dealing with a **merging of intangible qualities and when such definitions as change and inconsistency (Uranus) are put together with illusion and nebulousness (Neptune)**, it is difficult to analyze what is happening in concrete terms. This is demonstrated by the description of the condition of the United States' economy during the period of the conjunction, which is still in range as this is being written (1994).

If one were to take every report seriously, we would be totally confused about the state of the economy. Looking at various commentaries on the Stock Market for a single five-day period (from April 4-9, 1994), we can see evidence of this. On April 4, the Stock Market plummeted. This was reported by some as an indication that we were headed for a depression, by others as meaning inflation was taking over, and by still others that it meant nothing. The next day the market began to recover and this was analyzed by some as an indication of a strong economy, by others as a sign that we were really in for a depression because it moved up too quickly, and by still others that the whole thing meant nothing.

The Stock Market represents only a small segment of the material world, but it serves to illustrate one interpretation of the Uranus/Neptune conjunction—that it is **difficult, if not impossible, to project into the future with this combination**. I would add to this that anyone with Uranus aspecting Neptune in their natal charts, or experiencing transiting Uranus conjunct Neptune activating any planet in their horoscopes, should expect such experiences.

Investigating a combination as it is happening is an excellent way to learn about its meaning. Since the conjunction began to form, I have noticed certain feelings and attitudes (which could be ascribed to these planets) have been shared by many clients. I have known some of these clients for a long period of time and have noted that these are not their usual feelings and attitudes. What seems to be prevalent is anxiety, especially when these clients are trying to project plans into the future.

Time and again the anxiousness would emerge when the client was asking such questions, as "If I do such and so, what do you think will happen?" Or, "How will this action affect my life next year?" I found that my response was always, "You are getting too far ahead. Concentrate on the present." And every time it seemed to alleviate the fears. One client told me, that as a result of what I said, she would get up each morning asking herself what she could do in that day to feel as though she had achieved something. The tasks may be relatively small, but seeing accomplishment can make you feel better and help you to gain momentum.

You have the choice of feeling overwhelmed with what needs to be done and do nothing, or setting about **doing one small chore after another**. I have experienced both. The anxiety has not disappeared completely. At least once or twice a week I begin to feel the knots in the pit of the stomach and the wave of anxiety. But when this occurs I remind myself that I should concentrate on the present and I set out to do one task. What I have discovered is that in the course of that single day, I achieve much more than I had expected to and I find that I am asking myself why I had worried so much. This has become my pattern. I would like to say that I had totally eliminated the anxiety. Unfortunately, I have not, but these periods are becoming shorter and less frequent. And, my feelings of accomplishment are growing.

In dealing with those who were born in the 1990s with the conjunction, the preceding provides suggestions as to how you might bring out the best qualities of the person. Or if you were born with Uranus and Neptune in aspect to each other, particularly in hard aspect, you may find that the pattern described above is a natural part of your behavior.

You may already have learned that you **tend to feel overwhelmed or worry if you project too far ahead in your thinking**. This, of course, can be modified or accentuated by other factors in your horoscope, but the tendency is very likely there. All of us worry to some extent, but if it becomes a big problem, it can be relieved by concentrating on the present, as suggested above. You can still **plan for the future, but you should make the most of each day as well**. This is not a bad philosophy for anyone to follow, but it is particularly good for those with Uranus and Neptune in aspect in the natal chart.

Your imagination will probably be very strong with Neptune and when it is connected to Uranus may tend to **blow reality all**

out of proportion, as there is no grounding with either of these planets. So you may imagine the worst possible scenario, but when all the facts are in, you discover that the situation isn't nearly as bad as you thought it would be. This is not the only course of action, however.

Visualization falls under Neptune as well as worrying. And when you begin to worry about something, instead of stewing about it, you can **visualize the situation as you would like it to be**. This may be enough to, at least temporarily, allay your fears, but it may do more. The visualization may help you to better understand what is happening or change your attitude about it. Or you might even find that matters begin to evolve in the manner in which you would like them to go. Neptune may have to do with illusion, but dreams can come true under Neptune as well. Since impatience is connected with Uranus, you may not be able to visualize for extended periods of time. But even a few minutes here and there can produce results.

If we look at transiting Uranus activating natal Neptune or transiting Neptune aspecting natal Uranus we may see some of the same qualities because they are combined, but there are differences as well. Some of the same symptoms can appear when the two planets are interacting, but the main focus is not the same.

When transiting Neptune is aspecting natal Uranus, the emphasis is on Uranian qualities with the idea of letting go of some of them or rising above them. For example, you might have always been a rebel but during the time of this aspect, you could decide that instigating revolutions is not necessary to further a cause and you might try another approach. Or perhaps you are used to fighting for your personal independence and during the aspect you discover it is not essential to do so. You could realize that feeling inwardly free is more important than announcing it to the world. Or the exact opposite could be true. You might never have rebelled nor outwardly expressed your need for freedom. Then Neptune arrives and you find it necessary to rebel or declare your need for freedom.

The message is that you should **dissolve your old ways (Neptune) of expressing Uranian themes**, but the end result would depend upon your typical form of behavior. The trigger, however, might be the same. Possibly, you could discover spirituality, and your **newfound faith** might let you know that openly rebelling will bring results if you have never done this before. Or you might decide that you can **make changes without starting a revolution**, if you have always previously done battle. Or in the case of independence, you could begin to examine your need for independence because you feel that someone is trying to victimize you (Neptune). What you do with these circumstances will again depend upon your usual pattern of behavior.

Another possibility with transiting Neptune aspecting your natal Uranus, that is quite far removed from the foregoing, could be that you feel **inspired to be creative**. The opposite of that would be disappointment with your creativity. Your inspiration may be coming from a spirituality that is beyond you, or deep within you. Either way, it is difficult to judge what you are creating in a concrete way. Therefore, it is natural, when the planet connected with disappointment is aspecting the planet of creativity, to lack confidence in what you are doing. But material reality could be distorted during that time, so you should just keep expressing your creativity.

You might look to those who seem to approve of what you are doing for support and stimulation and you, personally, will probably be better able to judge your abilities after the aspect has passed. In the meantime, it keeps you busy and helps you to avoid some of the more negative manifestations. And you may even produce something worthwhile.

When transiting Uranus is activating natal Neptune, there may be a feeling of urgency, a need to take action. Whereas with Neptune, there can be dissatisfaction with present circumstances but a vagueness in regard to what is wrong. Therefore, it takes time to determine a course of action.

The overall message of transiting Uranus aspecting natal Neptune is that you should **change something Neptunian in your life**. There might be an impact on your spiritual life. If you are involved in an organized religion, you could express your individuality through it, or you might choose to leave the church to either **join a less restrictive religion** or follow your personal beliefs without being attached to any religion. Another possibility in regard to religion is that you might find yourself expressing your religious ideas vehemently to others, even to your own surprise.

If you have been feeling victimized by someone, you might choose to remove yourself from the relationship when Uranus is aspecting Neptune. You could just back away from it, but with Uranus you will probably have the strong urge to be openly defiant about it. The idea of making a stir will probably be much more appealing than quietly slipping away. Or you may not be feeling victimized yourself, but find that you are becoming aware of the victimization of others. At this time you might even **become involved with a cause** of this kind.

Since Neptune is also associated with **drugs and alcohol**, you should **be careful** with the use of these substances during the time of the transit. But it may also be a time when **you can suddenly become aware that you have a problem**. I have a client who had such an experience. It was during this aspect that she decided to join Alcoholics Anonymous, and has not only been "on the wagon" since that time, but also felt that her spiritual life has become enriched.

If none of the foregoing suggestions is appropriate for you, you can always choose to bring out your artistic qualities. Allow yourself to **explore what you can do in one of the arts**. You do not have to work at it. In fact, if you were to expend effort on your artistry for extended periods of time, the spontaneity that is part of the aspect would be missing. Your creativity should be expressed as the spirit moves you. Pressing too hard could make you tense and nervous. But allowing the energies to flow can release the tension.

Music particularly can play an important role during this aspect. Neptune is often associated with the "music of the spheres." You might discover that you have musical ability during the time that transiting Uranus aspects natal Neptune. But, if you do not have the inclination to explore the possibilities of this, just listening to music or dancing can be a release and relief. The movement of **dancing** can be very satisfying and there is nothing in the aspect that says you have to do it publicly.

The foregoing suggestions are only a few possibilities. You can perhaps find more suitable alternatives if you keep the principles connected with the planets in mind as you make choices. Concentrate on making the most of the moment. **Be spontaneous and go with urges to explore your spirituality, your creativity and**

your individuality. Anything that might help you to become more highly evolved or feel freer would be worth investigating. The only warning is that what you wish for should not be harmful to you or to anyone else.

Uranus and Pluto

Key words that are appropriate for a connection between Uranus and Pluto could be a **powerful (Pluto) revolution (Uranus)**. The last **conjunction** which occurred **between 1963 and 1968** certainly bore this out, at least in the United States. During those years, race relation problems in this country built to a crescendo and culminated with a series of riots in major cities across the country. (The events of this period were described in Chapter 6.) The sudden eruption of violence as evidenced by the riots easily fits into the astrological definitions of the planets; but when you consider the length of time Uranus and Pluto were conjunct before the riots took place, it is clear that the action was not immediate.

The slowness of the manifestation of this conjunction points up other qualities that we need to recognize about these planets. Although Pluto can represent eruption, upheaval, violence, it is also the planet of deep-probing analysis. And, we have mentioned time and again in this book, that Uranus has to define boundaries before it can move beyond them, or understand limitations before it can be free of them.

This may explain why those who were born from 1963-68 with the conjunction of Uranus and Pluto in their horoscopes, are not always starting revolutions. On the contrary, most seem to be quite conforming. You might attribute part of this behavior to the fact that the conjunction took place in Virgo, indicating that if these individuals are going to wage war, there had better be a good reason for it. But the nature of the planets themselves, as mentioned in the previous paragraph, has to be taken into consideration as well.

The detached, unemotional qualities of Uranus have already been discussed. Therefore, feelings are not likely to be considered when determining action to be taken. And Pluto certainly is not associated with nurturing. The movement taken under Pluto aspects is sometimes described as similar to a **bulldozer**. The driver of the bulldozer does not stop the machine and get down to remove a flower or save an animal. Anything in its path gets flattened and would not easily be restored to its previous state. And there seems to be a caution that those born with this conjunction seem to exhibit, as though they innately understand that action once taken cannot easily be undone.

If you have this conjunction, you should be able to relate to this concept and you may have already figured out how to make the best use of it. But for those of you who have not found an acceptable way to express the energies, one possibility seems very clear. When you want to make changes and assert your power, you have to make sure you are taking the best path. In this particular case, with both planets in the sign of Virgo, **the more facts you unearth, the more likely you will be satisfied with the end result**. Therefore, you need to investigate options. You do not have to do this all at once. You may gather a little information and test your findings with a small action because trying it out before you make sweeping changes will ease the transition.

There may be times when you feel impatient. Those are the best times to take small actions to make yourself less nervous and to keep you from erratically creating untenable situations. By dividing up the material you are investigating into small units you can take action and analyze results on each unit before moving on to the next. Because you are frequently evaluating your findings you may not be quite so impatient and eventually the dramatic changes will take place.

For those of you who do not have the Uranus/Pluto conjunction but interact with people who do, you would be wise when jointly making decisions that involve change and/or power to present the facts as clearly and succinctly as possible. The better you do this, the faster the action. The individual with the Uranus/Pluto has to understand and analyze the situation and this takes time; but, once a decision is made, the action itself will occur quickly and, as with the riots of 1967, you will be aware that action has taken place.

There are, of course, other aspects between these two planets that can appear in a horoscope. The particular signs and aspect will modify the definition, but the principles will remain the same.

A powerful revolution that will result in great change is possible with this combination, but the momentum will grow gradually and the action will probably not be as spontaneous as it appears.

In looking at transiting Pluto aspecting natal Uranus or transiting Uranus aspecting natal Pluto, you could still be dealing with power and great change, but there are other possibilities as well. If transiting Pluto is aspecting natal Uranus, you could covertly express your individuality or independence. You may feel forced to do this because someone is ordering you around, or circumstances are making you feel oppressed. If you are involved with people you cannot put in their place, or situations that you feel unable to change, finding a way to feel free without confronting the condition may be what you should do. Escaping temporarily from such circumstances may make you better able to tolerate them when you return.

When **Uranus is aspecting natal Pluto, you might be called upon to assert power suddenly, or you might relinquish power**. I have a client who had mother-in-law problems. From the moment she married her husband, there seemed to be a constant battle about whom he belonged to. During a period when they went to spend a week with his parents, my client was in the midst of transiting Uranus squaring her natal Pluto and she was sure it was going to mean another skirmish. But what happened instead was that, as her mother-in-law was playing her manipulative games, my client suddenly decided she did not want to play the game any more and that she did not need to. In a flash she realized that her husband lived with her and she did not have to worry about her mother-in-law usurping her power. When she ceased to respond, her mother-in-law no longer had an adversary to battle with. At that point, their relationship began to take a different turn. It still is not ideal, but it is better than it was.

When transiting Uranus begins aspecting natal Pluto you don't have to wait for an event or confrontation, to start to examine how and where you can use your power. You could decide as my client did, that it is time to change the way you assert yourself and possibly **let go of outdated issues involving power**.

CHAPTER 14

URANUS IN CHARTS

It is doubtful that any astrologer would start an interpretation of a natal horoscope by discussing Uranus. In fact, in forming an overview of someone, this planet may sometimes be ignored. As mentioned earlier in this book, there are those who believe that this transpersonal planet may provide information about the masses, but has little significance for individuals. Hopefully the preceding pages have helped to convince you otherwise. To further convince you, let us look at horoscopes of people from two perspectives. One is **to see what Uranus can tell us about these people**. The second is to look at the individuals through the dynamics of their horoscopes, using **transits and progressions** to see what they can tell us about the planet Uranus. Since some of the qualities and types of issues that Uranus represents are independence, revolutionary tendencies, sudden change and creativity, we should keep these definitions in mind along with any others we might use, as we look to this planet in a personal horoscope.

We will begin to investigate Uranus in a natal chart by interpreting it in terms of the **house in which it is placed**. This will be an area in which the **native is likely to break with tradition**. It will also indicate where Uranian issues will be expressed most directly. The **sign in which Uranus is posited** will then be added to explain **how the individual deals with Uranian matters**. The characteristics indicated by the sign will, of course, be shared by all people born within an eight-year period.

But the house will not be shared by everyone with Uranus in a particular sign, so we have begun to personalize the planet. Then when the aspects of Uranus to other planets and points in the horoscope are discussed, an even more distinct pattern of how an individual will express the qualities associated with the planet Uranus is formed. Let us first examine the horoscope of John F. Kennedy.

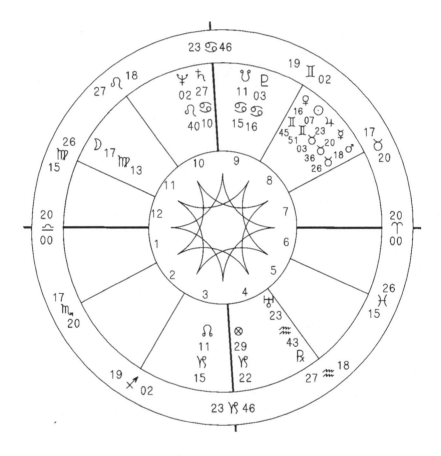

John F. Kennedy

May 29, 1917 15h0m0s EST
Brookline, MA 42N19 71W07

John Fitzgerald Kennedy

Kennedy was selected because his life history is well known, and also because he was used in this book as an example for the Uranus in Leo period in the history of the 20th century. Thus we can look at Uranus from both the mundane and personal perspective. As President of the United States during the time that Uranus was in Leo he epitomized, for the nation, the meaning of the planet in that sign. However, the way in which he personally dealt with change and his own revolutionary tendencies were quite different from the way the world experienced his actions in regard to these and other Uranian matters.

Natal Placement of Uranus

He was born with **Uranus in the fourth house**. Therefore, he could have had an unusual home life, or there could have been frequent changes, or freedom, in the home. There has been enough written about, and shown in films and on television, concerning the Kennedy household as JFK was growing up, to know that it fit the placement of Uranus in the fourth house. They **moved a great deal**. And their **lifestyle was quite different** from the typical American household of the period in which John Kennedy was raised. It was dissimilar as well, from their ancestors (also the fourth house) who came from Ireland.

There may have been some rules and regulations placed upon the nine Kennedy children, but from all reports, their **creativity and individuality were encouraged**, and the household itself was quite unusual. Then, there are stories from his adult life, showing that he seemed to continue the cultivation of freedom in the home. For example, during the short time he was in the White House it was supposedly not unheard of for his daughter Caroline to feel free enough to walk into an official meeting her father was having.

Another meaning of the fourth house, which is appropriate for John Kennedy since he was President of the United States, is the "homeland," since he officially represented his country. With Uranus in the fourth house it might be expected that he would have **revolutionary ideas in regard to running the United States**. His program, called the "New Frontier," certainly suggests the idea of exploration, which is an excellent definition of Uranus. It was during his time in office, and credited to his term, that the Peace Corps, epitomizing the **humanitarian principles** we ascribe to Uranus, was created.

He was born with **Uranus in the sign of Aquarius** during the period in which Woodrow Wilson was in the White House. And Kennedy personally reflected the visionary qualities associated with Wilson's handling of the Presidency. As explained earlier in regard to the Uranus in Aquarius period in American history, Wilson's visionary ideas did not all come to fruition. The concepts are clear on a mental level with Aquarius, but implementation does not easily come about with that sign.

Although the Peace Corps did materialize during Kennedy's term in office, other ideas that he had did not take shape until after his death. With Uranus in Leo during his Presidency, Kennedy was able to verbalize his dreams, and the public could understand them. But it was not so clear as to how he might make them manifest. As a matter of fact, it was during Johnson's term — when Uranus was in Virgo—that many of these goals were accomplished.

Aspects to Uranus

As we add the aspects to Uranus in Kennedy's chart, we can get more insight into his use of the energies associated with that planet. He was born with a square between his Ascendant and his Midheaven which indicates that there was a potential conflict between his public image (Midheaven) and his personal self (Ascendant). When we note that **Uranus trines the Ascendant** and almost exactly **quincunxes his Midheaven**, the implications become more specific. He could express his individuality and creativity without effort (Uranus trine the Ascendant), but when it was not favorably accepted by the outer world (Midheaven) he would have to make an adjustment (quincunx) spontaneously (Uranus to the Midheaven).

His **Uranus also squares a conjunction of Mars, Mercury and Jupiter**. The triple conjunction certainly indicates his ability to communicate and this was evident in both his speaking and his writing. He was called the "media" President and often demonstrated verbal facility. He also won the Pulitzer Prize for his writing. Since Mercury represents thinking as well as writing and speaking, and it is straddled by Jupiter on one side and Mars on the other, he obviously **had big ideas, could create grand plans and was able to make them sound very plausible** (Mars, Mercury and Jupiter are all in Taurus).

Now add the square of Uranus to these three planets and a more elaborate picture is formed. Uranus/Mars would mean that Kennedy was **impatient** and could act impulsively. Uranus/Mercury, indicating a **quick thinker**, repeats the message of spontaneous communications, and might also mean that he **sometimes would speak without thinking**. Uranus/Jupiter is the **explorer or risk-taker**.

Because Mercury falls in the midpoint of the Mars and Jupiter, the impulsive actions and possible risk-taking would probably fall more often in the realm of speaking or writing rather than physical activity. All of this is indicative of someone who was **forward-thinking, but also might mean that he could sometimes be a bit rash**. Since Uranus squares these planets he would probably periodically have to deal with and resolve such issues.

The triple conjunction of Mars, Mercury and Jupiter aspects the Midheaven and Ascendant as well. Thus the meaning of the conjunction is projected into his public image (Midheaven) and his personality (Ascendant). The conjunction sextiles the Midheaven so that the energies connected with Mars, Mercury and Jupiter are easily directed in his career. There may even be help from others in expressing them. However, the conjunction quincunxes the Ascendant inferring that there would sometimes need to be an adjustment in his personal life in terms of actions (Mars) he might take, what he may say or write (Mercury), and how he tries to grow and develop (Jupiter). This may be borne out by his alleged personal indiscretions which were brought to light long after his death.

All of these planets and points fall between 18 and 23 degrees and the aspects they form are easily seen. Other less visible aspects could be brought into the picture, but the ones already mentioned and described should be enough to explain how Uranus was expressed in his life. They also can provide a basis for examining dynamic techniques.

Transits and Progressions

If transits and progressions were investigated with a traditional approach, all the planets would be looked at. Periods of high activity could be noted because of the number of aspects formed. Also, types of activities during any given period would be determined by the particular planets that were prominent during that time. But the emphasis in this book is only on Uranus, and therefore, the approach has to be different.

Just as Uranus describes only part of the personality in the natal horoscope, in the dynamic techniques it will not account for all that is occurring in someone's life at any given time. The drives ascribed to it may contribute to certain events or attitudes, but any specific event will most likely reflect the issues associated with a number of planets. And there may at times be important events in which Uranus is not prominent at all. Therefore, it will be most efficient to present a chronological listing of events in Kennedy's life. Then, as we move from event to event, we can look for aspects from all transiting planets to natal Uranus and aspects from transiting Uranus to all natal planets and points for the various events and see the role (if any) that Uranus played.

List of Events in John F. Kennedy's Life

Born May 29, 1917
Entered Princeton Fall, 1935
Withdrew because of illness after Christmas
Entered Harvard Fall 1936
Graduated Harvard cum laude 1940
Enlisted in the United States Navy 1941
PT boat which he commanded was destroyed
        by the Japanese August 2, 1943
Through his efforts, he and his crew were rescued
        August 7
Elected to the House of Representatives
        November, 1946
Elected to the Senate November, 1952
Married Jacqueline Bouvier September 12, 1953
Won the Pulitzer Prize May 6, 1957
Elected President November 8, 1960
Inaugurated President January 20, 1961
Set up Peace Corps March 1, 1961
Bay of Pigs Invasion April 17, 1961
Cuban Missile crisis October, 1962
Assassinated November 22, 1963

As the events of Kennedy's life are viewed, **concentration will be on sudden change, creativity, independence, and revolutionary qualities connected with the events**. Uranus in transit will be examined in terms of its connection to the natal chart. Other planets in transit aspecting the complex configuration containing Uranus, which was mentioned above, will also be discussed when they contribute to understanding the meaning of the planet Uranus.

If we were to use minor aspects and midpoints, the role of Uranus could undoubtedly be found in every event. But because we are searching for simple and direct information about the planet we are looking only at the obvious aspects (the 30° aspects).

In the first event, which was actually a series of events, **he entered Princeton** in the fall of 1935 and withdrew after Christmas because of illness. On September 1, 1935 **transiting Uranus** was retrograde at 5 Taurus 20. It had turned retrograde on August 11 at 5 Taurus 31. Therefore it had already **squared his natal Neptune and sextiled his natal Pluto**. Both aspects recurred in November. Transiting Uranus sextiling natal Pluto might indicate sudden transformation or getting in touch with power issues. The square to Neptune could mean being faced with the need to suddenly let go of something, or indicate awakening in terms of spirituality. The combination of transiting Uranus square natal Neptune might also manifest as the dissipation of energy or feelings of nervousness and discomfort. The square of transiting Uranus to his natal Neptune could show that he was prone to illness at that time. Had Kennedy known astrology he might have been able to avert the physical problems by making use of this transit in other ways.

Transiting Uranus went direct on January 10, 1936 at 1 Taurus 33 so that it formed the same aspects for a third time, and then moved on. By April the aspects had separated, his health improved, and **he entered Harvard in the fall of 1936**.

Let us return to the transits for the fall of 1935. In early September of that year, transiting Mars was in Scorpio and activated all the planets and points between 18 and 23 degrees in Kennedy's horoscope. **Mars squared Uranus; opposed Mars, Mercury and Jupiter; semisextiled the Ascendant; and trined the MC**.

Transiting Jupiter, too, was in Scorpio at that time and, between mid-September and early October, formed the same aspects as Mars with the aforementioned natal planets and points.

The two transiting planets never conjuncted each other. In fact, the Mars aspects were complete before the Jupiter aspects began. This would indicate, in general, that action and invigoration (Mars) took place before expansion (Jupiter) began. A proposed scenario could be, that with Mars square Uranus, he suddenly took the initiative and went to college. The opposition of transiting Mars to natal Mars, Mercury and Jupiter would indicate a great deal of activity both physical and mental. However, neither transiting Mars square Uranus nor opposition the triple conjunction of Mars, Mercury and Jupiter would suggest concentrated effort. The **attention span would probably be short**. Or perhaps there were a number of different kinds of activities that had to be addressed so that there would not be time to concentrate on any for too long. The trine of transiting Mars to the MC would indicate that his public image moved ahead easily, but a personal adjustment (Mars semisextile Ascendant) was necessary.

In a matter of days after Mars formed the aforementioned aspects, Jupiter moved to the same point in the zodiac and activated all the planets and points involved, in the same manner. The first aspect from Jupiter was exact on the 8th of September and the last aspect was exact on the 15th of October, whereas, the first Mars aspect was exact on August 28 and the last was exact on the 7th of September.

The exactness of an aspect does not indicate that everything that is going to occur in regard to the combination will happen on that day. **Events may manifest at any time during the period in which the transiting planet approaches the natal planet until it separates from it.** The specific dates are mentioned solely to show that Mars was moving more rapidly than Jupiter, so its impact would likely be more short-term.

The principles involved with Mars and Jupiter are different as well. Mars energizes and triggers. Jupiter expands and develops. Even without astrology, the sequence makes sense in light of what was happening in his life. First he went to college (Mars) and then he expanded his horizons (Jupiter). We could, therefore, talk about how he expanded his individuality (Uranus), his mind or communications (Mercury), his intellectual development or knowledge (Jupiter), and both his public and private image (MC and Ascendant) through the particular aspects.

Before moving on, however, it is important to recognize that **with both Mars and Jupiter it is possible to spread yourself too thin**. Because of the high physical energy possible with Mars, you can keep taking action until you feel exhausted. And with Jupiter you have so many ideas that you can undertake more than you can accomplish, and stop in your tracks because you feel overwhelmed. Add all of this to the transiting Uranus square his natal

Neptune, and we have a story that possibly explains why he became ill and withdrew from Princeton after Christmas.

The timing of the withdrawal might be further explained by transiting Mars which moved through those critical degrees of 18-23 from the end of December to early January. Mars was in the sign of Aquarius then, so that the specific aspects were different from those that occurred in September.

Following his four years at Harvard, **he enlisted in the United States Navy in 1941.** During 1941 **Uranus was in Taurus**. As the year began, it was retrograde at 22°. It had already squared his natal Uranus twice and during 1941 squared it one more time. He was experiencing the urge to be independent and free himself from previous ties. At the same time, however, Saturn was also in Taurus and during the course of the year conjoined Uranus in the sky and, therefore, squared his natal Uranus as well. This may explain why he left his old life behind (**Uranus square Uranus**), but also why he aligned himself with the established, rigid structure (**Saturn square Uranus**) of the United States Navy.

Transiting Uranus also conjoined his Jupiter and sextiled his MC. This could mean that he was being adventurous (Uranus conjunct Jupiter), and he was easily taking on a new public image (Uranus sextile MC). That is the total picture of transiting Uranus for the year in regard to the planets and points between 18° and 23° in his natal chart because Uranus only moved between 22° Taurus and 0° Gemini and, therefore, had already passed his natal Mars, Mercury and Ascendant.

Saturn, on the other hand, did conjunct Mars and Mercury and quincunxed his Ascendant. All of this would indicate that although he seemed to take sudden, drastic action he probably seriously (Saturn) thought about (conjunct Mercury) what he was doing (conjunct Mars) and, with effort (Saturn), made his personal (Ascendant) adjustments (quincunx). We might add, with the Saturn conjunct Mars, that he probably had to make specific plans before he could take definitive action.

The next event was the **destruction of his PT boat by the Japanese on August 2, 1943,** and we would expect Uranus to be prominent. It certainly was, after all, a sudden, unexpected event. And **Uranus, by transit on that day, was at 8° Gemini 3"**—

12 minutes from an exact conjunction to his Sun. Jupiter, on that day, was at 7° Leo, moving to a sextile with the Sun. On August 7, the day of the rescue it was at 8° Leo 8', just past the exact sextile, which astrologically can explain the rescue. Transiting Pluto was within 1° of a conjunction with transiting Jupiter. So it, too, was in range of a sextile with the Sun and further contributes to the astrological explanation of the rescue. The Pluto aspect might also be interpreted as contributing to a personal transformation, which is a logical possibility after experiencing such a devastating event.

To further describe what occurred, transiting Mars and Saturn were both activating those planets and points between 18° and 23° in the natal chart. Mars was in Taurus. On the day of the disaster it was at 16°, and the day of the rescue it was at about 20°. So in that short five-day period it conjoined natal Mars (18 Taurus 26) and quincunxed the Ascendant (20 Libra 00). **Transiting Mars conjunct natal Mars would indicate the need to take action** and it was supposedly through Kennedy's efforts that they were saved. And transiting Mars quincunx natal Ascendant shows the potential for personal adjustment which in such circumstances would be necessary.

At the time of the rescue, Mars was conjoining his natal Mercury (20 Taurus 36), and assuredly there were active communications. Then it moved on to a conjunction with his Jupiter, square his Uranus, and sextile his MC. The **Uranus square itself would describe the abrupt change that occurred both with the sinking of the ship and the rescue**. The other aspects could be connected with his being saved (Mars to Jupiter) and recognition and praise of his efforts (Mars sextile MC). Slow-moving Saturn moved from 22 Gemini 36 to 25 Gemini 13 in the month of August, 1943. It therefore was semisextile his Jupiter and trine his Uranus, perhaps stating that he had some control over his circumstances. Transiting Saturn also semisextiled his MC indicating that he would have to work on readjusting to the outer world.

The next events listed in John Kennedy's chronology are his **election to the House of Representatives in 1946 and his election to the Senate in 1952**. Even though the election takes place on a particular day and the results may be a surprise, there is a great deal of groundwork that needs to be laid and the entire process is slow and plodding. Therefore, we would not look upon these events as Uranian.

So let us move on to his wedding with Jacqueline Bouvier on September 12, 1953. By transit, **all six of the outer planets** (Mars, Jupiter, Saturn, Uranus, Neptune and Pluto) **had activated, or were about to activate (using a 1° orb), Kennedy's natal Uranus, Jupiter and MC**. Transiting Jupiter was at about 24° of Gemini, trining natal Uranus, semisextiling natal Jupiter, and semisextiling natal MC. Transiting Neptune was at 22° Libra, trining transiting Jupiter, and forming a grand trine to natal Uranus. So we might say that major changes with emphasis on his freedom, creativity, etc. could have been occurring without effort (grand trine) at that time. Neptune was also quincunxing natal Jupiter and squaring natal MC, indicating that adjustment (quincunx) and effort (square) were required in dissolving (Neptune) patterns associated with his development (Jupiter) and public image (MC).

Transiting Saturn was at 24° Libra at the beginning of September, within range of a conjunction with transiting Neptune. This suggests that, at the same time that **patterns were dissolved (Neptune), new structures (Saturn) were being formed**. With Saturn as part of the grand trine, we add parameters and stability to the changes taking place—more or less establishing ground rules for the changes. This would be true for his sense of personal development (Ascendant) and his public image (MC) as well.

To complete the picture of the period, transiting Pluto was at 23° Leo, and transiting **Mars conjoined transiting Pluto** at the beginning of the month. These planets then **opposed natal Uranus, squared natal Jupiter and semisextiled natal MC**. Obviously, there was so much activation of the horoscope that something of monumental importance would certainly have been happening at that time. So it seems evident that **his marriage was important in reshaping his life**.

The next event listed in Kennedy's chronology is the **Pulitzer Prize** which he was awarded in 1957. The presentation was made on May 6, 1957, but the process of being selected for such an honor takes much longer than a single day, or even a single month. First the book had to be written. Next it had to be submitted as an entry for the award, and then it had to be chosen. Therefore, it is difficult to pick a short time frame that should be examined in terms of transits. It might be mentioned, however, that at the time the award was presented transiting **Uranus was semisextiling natal Pluto**, indicating a possible minor adjustment (semisextile) with power or transformation (natal Pluto) coming suddenly (Uranus).

Also, **transiting Jupiter** was retrograde at 22 Virgo. It had already **trined natal Jupiter, sextiled natal MC and quincunxed natal Uranus**, and in June it activated those points again. Therefore, he grew and developed quite easily (Jupiter trine Jupiter) and his public image was enhanced by others (Jupiter sextile the MC). And there must have been some adjustment in his sense of freedom or individuality (Jupiter quincunx Uranus). But aside from showing the validity of transits, this event is a confirmation of the potential in the natal chart for talent in writing. The conjunction of Mars, Mercury and Jupiter alone would be enough to show this ability, but it is expanded upon by the aspects formed by this triple conjunction to other planets and points in the chart.

The next event was Kennedy's **election to the Presidency** on November 8, 1960. As with the other political offices to which he was previously elected, as well as the award of the Pulitzer prize, this process was a slow-moving one. But it is worth mentioning that, at the time of the election, transiting Mars was at about 17° Cancer. It went retrograde at 18 Cancer 39 on November 20, 13 minutes of an exact sextile to natal Mars. It might also be noted that from November 5-10, **Venus moved from 18-23° Sagittarius forming a yod with the triple conjunction and MC of the natal chart and sextiling both his natal Ascendant and his natal Uranus**. Although transiting Venus is not usually connected with milestone events, it does show that the pleasure principle was active in his life during this week, creating a change (quincunx) and making him feel good about himself (sextile Ascendant) and his individuality (sextile Uranus).

Taking the broader view of the period, during the preceding months and the following ones, transiting **Uranus opposed its natal position three times**. This is, of course, part of the Mid-Life

Transition for everyone. And Kennedy certainly had a major change in his life during that time.

In February, after entering office, Uranus moved back to 23° of Leo (for the second time). There was also a Jupiter/Saturn conjunction in the sky during that month at around 23° Capricorn (semisextiling natal Uranus, trining natal Jupiter and opposing natal MC). A plausible story line, with Kennedy's situation and the planets involved, would be that there were definite and sudden changes that had to be handled (Uranus opposition Uranus). The principles of expansion (Jupiter) and contraction (Saturn) were having an impact on his individuality (semisextile natal Uranus), his public image (opposition natal MC) and his own ideas of development and expansion (trine natal Jupiter).

The next events were the **starting of the Peace Corps on March 1, 1961 and the Bay of Pigs disaster on April 17** of the same year. **Transiting Uranus did not make its final pass opposite natal Uranus, square natal Jupiter, and semisextile natal MC until July of 1961**. Therefore, the contrasting events of the Peace Corps which brought him praise, and the Bay of Pigs episode which brought him criticism, fit easily into the plot of the period. The transiting Jupiter and Saturn conjunction seemed to be still in range, although it had passed 23°. The Jupiter/Saturn contribution to the scenario was still in effect, and slow-moving planets in transit frequently last longer than the traditional 1° orb. There is another factor that can be added in April, however. On the day of the Bay of Pigs invasion, transiting Mars was at 20° Cancer, sextiling natal Mercury at 20° Taurus. It had just sextiled natal Mars and squared natal Ascendant. And **by April 25, Mars had made a quincunx to natal Uranus, a sextile to natal Jupiter, and a conjunction to natal MC**. All of this describes the potential of taking the initiative, or committing an act of aggression. In the previous month, however, when the Peace Corps was established, the aspects were not in range.

It is interesting to note that in September 1962, the month prior to the **Cuban Missile Crisis** (another incident which had to do with assertiveness, taking the initiative, etc.), **transiting Mars had returned to the sign of Cancer and again formed all those aspects mentioned above**. Added to that, transiting **Uranus was opposing transiting Jupiter** (the risk-taking combination) and

Jupiter was trining Kennedy's natal Pluto while Uranus was sextiling natal Pluto. Thus, the astrological foundation is in place for what ensued.

I am hesitant to discuss the final event, his assassination, because sometimes people associate death with certain aspects when they are present at the time of someone's demise. Then when these aspects appear in their own charts they tend to worry unnecessarily. So let me emphasize that **there are always many different ways in which any aspect can be manifested**.

In Kennedy's case, in November 1963, transiting Mars was in Sagittarius. During the course of the month it quincunxed natal Mars, Mercury and Jupiter in his eighth house (which in astrology is the house of death, among other things). Guns fall under Mars and the quincunx is an aspect of change and adjustment. **Transiting Mars also formed a sextile with Kennedy's Uranus** (indicating the possible suddenness of the event). And, finally, Mars quincunxed his MC. Thus Mars formed a yod with natal MC on one side and the natal Mars/Mercury/Jupiter conjunction on the other. With the focus on Mars (the gun) a dramatic change could occur (Yod) that involved his public image (MC) and his physical vigor (natal Mars), his communications (natal Mercury) and his personal development (natal Jupiter). Since the triple conjunction is in the eighth house, the adjustment could have occurred through death. However, the dramatic change could have been a personal transformation (another meaning of the eighth house) as a result of his taking the initiative (transiting Mars). And it would be possible to expand on this story line just as fully as the one describing his death.

The progressions are not usually as dramatic as the transits because progressed planets and points move so slowly. Sometimes they are evident in events, but they work more consistently when used to explain how a life evolves. This is particularly true when a very slow-moving planet such as Uranus is investigated. In the birth chart of John F. Kennedy, **Uranus was retrograde at 23 Aquarius 43**. At his death it was still retrograde and had moved only to 22 Aquarius 54. Therefore, **during his entire lifetime it moved only 49 minutes** so it **always formed a fairly close square to his Jupiter and a quincunx to his MC**. At birth, Uranus was already passed the quincunx with the MC but **in 1957/58 the square to Jupiter became exact**. That was the year (1957) in which he received the **Pulitzer Prize** which certainly epitomized the "sudden windfall" concept of that combination and other facets of the combination might have been obvious in other events of the period.

As already stated, however, it is the evolutionary process that is unique to progressions and, therefore, is a more important element of progressions than events. An excellent illustration in Kennedy's life involving Uranus was the **opposition of progressed Uranus to progressed MC**, which could indicate a possible change in career direction. It was well-known that his brother Joe was supposed to have been the politician in the family. JFK did not have aspirations in that direction, but when his brother died, he moved into the role. And it is interesting to note that when he was elected to the House of Representatives in 1946 the progressed MC was about 2° away from, and approaching, both natal and progressed Uranus. In 1952, when he was elected to the Senate, his progressed MC was 4° past his natal and progressed Uranus. This undoubtedly means that somewhere in that six-year period, the change in career direction actually concretized—not only in terms of events—but also as part of his internal development.

We might look further into progressions by examining the progressed Moon which, of course, moves most rapidly. It completes a cycle every 27+ years and would have been in its second cycle at the time of Kennedy's death at the age of 46. Therefore, we could note its position in regard to Uranus at the time of the events given and the reader might do that. For our purposes, the preceding example should be enough to show how progressions can be connected both with events and with the evolution of an individual.

Another Horoscope

Studying a life after the fact can explain how a planet materialized for a specific individual, and this is an excellent tool for learning. But we want to understand and make use of the energies connected with the planets before events occur rather than after.

As we **concentrate on utilizing Uranus in our lives rather than just gathering information** about it, we can start in the same manner as we did with Kennedy. As the horoscope of an individual is examined, the house and sign in which Uranus is placed will still describe where the individual will break with tradition and where one can most directly express individuality and freedom. The aspects to Uranus add more information.

Whether you are dealing with an adult or a child, both positive and negative possibilities should be determined. If a child's horoscope is being interpreted, it can be discussed with the parents as to how they might try to help the child accentuate the more desirable qualities. With an adult, you can still talk about alternatives. And, although you may not have a biography as you would with a famous person, you have a certain advantage over interpreting a child's chart. The advantage is that you can ask how Uranus has been expressed thus far in the life. This will provide clues on how the individual will most likely react to transits and progressions to and from Uranus. If Uranus has been experienced in a negative manner you can also talk about how to work on changing it.

The chart of the individual given below shows **Uranus conjunct the Ascendant on the twelfth house side**. With Uranus within a degree and one-half of the Ascendant, this person could be strongly individualistic, but that would be more likely if Uranus were on the first house side rather than on the twelfth house side. As stated earlier in this book, if a person with a twelfth house Uranus were encouraged to develop creativity, express individuality and be independent in childhood, these qualities will be easily manifested in adulthood. If they were not, Uranus themes can present problems as the person grows up.

Asking the client whether or not she was encouraged to express her individuality as she was growing up, is certainly the easiest way to determine if her Uranus was allowed a direct outlet. She was asked this question. Her answer was no, she was not encouraged to express her individuality. She had been born in Germany of Russian parents who were very strict in raising their children. The woman felt that she had little freedom when she was a child.

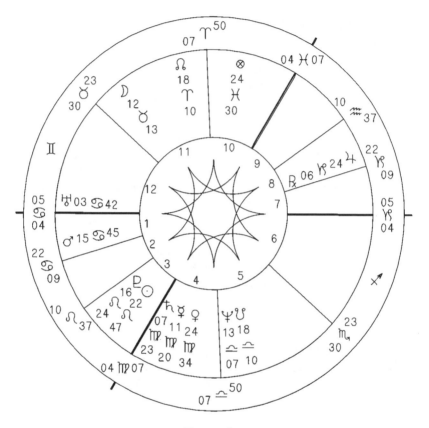

Female
August 16, 1949   2h2m0s METD
48N46 11W27

It can be assumed, with her **Uranus in Cancer**, that she probably is best able to express her individuality or revolutionary tendencies when she is taking care of the needs of others. This would have been true whether or not her early environment had been restricted. Had she been allowed or encouraged to express her Uranian qualities, she would still most likely have utilized these traits for others. The main differences would be that she would have felt more comfortable rebelling in a freer environment and would have done it spontaneously. Because of the way she was raised, she might often have suppressed urges to be different, but when she showed this side of herself, she probably felt some guilt and would have to make excuses if she were challenged. Saying she was doing it for someone else, or was asked to do it by another person, were probably the reasons she gave if she were forced to explain her Uranian actions.

The **sextile of Uranus to Saturn** makes the possibility of this scenario even stronger. If she were doing her duty, it might be acceptable to use unusual means. This is certainly a good way to use the combination of Uranus and Saturn. The sextile would mean that others may have helped her to do this or supported her behavior, and this should be considered and made use of when Uranian issues arise.

Since she was not encouraged to express her rebelliousness or individuality by her parents, this sextile might also add to the problems that need to be worked out in her lifetime. One such problem which was already mentioned could be feeling guilt for expressing her individuality. Performing her duty before she does something for herself might be one way in which she could alleviate the guilt. Developing a pattern of behavior which she can call upon each time issues of this sort arise, might help her to develop and better express her Uranian qualities.

She also has a **trine between natal Uranus and natal MC**. Positively, this would mean that if the world approved of her Uranian behavior, it would become easier and easier to express. But conversely, disapproval of her displaying her individuality—showing how she is different from others—could cause her to crawl more deeply into her shell. One can visualize someone triggering this reaction by pointing out that others do not behave the way she is behaving. And this individual, with the sensitive sign of Cancer rising and a fixed Leo Sun and Taurus Moon, who would not make changes so easily anyway, could find that shell very appealing.

The final Ptolemaic aspect involving Uranus in this individual's horoscope is the already mentioned **conjunction to the Ascendant** from the twelfth house side. This would indicate that Uranian qualities should be an integral part of her total personality that she should express in the outer world. Since this part of her was not encouraged as she was growing up, it would definitely be something that would have to be worked on in adulthood.

An astrologer working with this person could share the preceding ideas and suggestions and discuss further ways to make use of Uranus by house, sign and aspect. Since she should break with tradition in the twelfth house, she might become involved in **unusual mystical traditions** and/or if she wishes to probe into the subconscious, she might use a kind of **therapy that is out of the ordinary**. Her study of a mystical tradition or involvement in therapy should be **thorough and as practical as possible (Uranus sextile Saturn)**. And with this aspect, even if her head is in the clouds, her feet should be planted on the ground. Or perhaps, she can find a **practical way to express what she learns, possibly by helping others (Uranus in Cancer)**. To complete the picture, she should **surround herself with those in the outer world who can support and approve of her behavior (Uranus trine the MC)**. Then she will have an audience whom she can allow to see the **nonconforming part of her (Uranus conjunct the Ascendant) without serious repercussions**. There are, of course, other ways to utilize the Uranian energies in the life, but the principles of Uranus by house, sign and aspect should be kept in mind as choices are made.

This combination should come into play particularly when natal Uranus is activated by any transit and/or progression. Chances are these planets and points would often act as a unit anyway, even without awareness. But the more conscious the native is of this pattern (and others) in the chart, the more likely it will be that she would use the energies in an appropriate way, and feel more in control of her own destiny.

This pattern also could be activated to some extent when transiting Uranus is aspecting any facet of the chart. The transit could indicate, among other things, that she was restless, had the urge to free herself, or wanted to change something in her life in terms of the planet or point being activated in her natal chart. Therefore, she might incorporate the planet in her natal chart that could help her move quickly—or describe how she could change—the planet Uranus. Then, as she makes use of Uranus in her natal chart, the entire behavioral pattern connected with natal Uranus would come to the fore.

The more awareness she has about the pattern, the more effective she can be in directing it. For example, when Uranus by transit comes up to oppose her Sun, she might want new ways to gratify her ego, or have the urge to change her life direction. As she is trying to make changes she might feel restricted by the caution (and sometimes restriction) Saturn represents. She might then give up and decide it is easier to forego happiness and not make changes. Or she could work on understanding why she is feeling limited, so that she can determine the best way to move on. Or she might perform some duty so that she can permit herself to make some changes without feeling guilt or fear.

Everyone has Uranus in the natal chart. This means that we should **all break with tradition at some time and in some place in our lives**. We should also **express our individuality, our independence, and our creativity** from time to time. Natal Uranus and the dynamic techniques can help us to know when and how. The better we grasp the implications, the more effectively we can use what is there.

CHAPTER 15

CONCLUSION

The Uranus journey is now complete, and it is time to attempt briefly to place this planet into perspective. Uranus is only one part of an individual's character that is described by the natal chart, and it is important to incorporate Uranus into the total picture.

Since a symbol of the universe is used to depict who we are, and the universe moves smoothly forward with few collisions, we should each be able to live in harmony within ourselves. Problems do arise in our lives, however. Sometimes these difficulties come from imbalance. **We may tend to accentuate certain facets of ourselves, possibly because they are simple to deal with, and try to disregard other parts** because they do not move so easily. When we are aware of symptoms of imbalance, we can concentrate on restoring harmony by accentuating what has been excluded, and cutting back on what has been over-emphasized.

There is the possibility, however, that you been balancing the various facets of yourself very well and are living in relative serenity. When your peace is interrupted you might either wonder what you did wrong, or why someone else is causing you this discomfort or unpleasantness. The answer to these concerns is that **life is dynamic and we have to move with the times**. Since each planet has many different meanings, as life changes you may have to examine other ways that may be more appropriate for you to express the energies at that particular time. So whether your new discomfort comes from within or it seems to be created by someone else, the message is that no one is to blame. It is just time to make changes.

Usually **transits and/or progressions will give you the message that it is time to make changes**. They will explain what is happening and offer clues as to how you can improve conditions. If you note the messages in advance, you might be able to make changes before difficulties force you to do so. In this way you could avoid some problems; but even if it does not work to perfection, you will at least be doing something other than just worrying about what will befall you.

In this book, Uranian themes in the natal chart have been discussed through the signs, houses and aspects. The dynamic possibilities of the planet were covered through progressions and transits. It is now up to you to use the principles that have been presented.

If you believe that you are uncreative, deprived of freedom or personally stifled, you should look for ways bring out these Uranian qualities. If you feel too detached or too free, you might want to accentuate energies associated with other planets to modify and blend with Uranus. In other words, you may want to **explore new ways to incorporate Uranus in your life**. So the journey of this book is over, but your journey may be just beginning.

APPENDIX

EXAMPLES OF URANUS IN HOUSES, SIGNS, ASPECTS

All examples listed in this Appendix come from *The American Book of Charts* or *Profiles of Women* by Lois M. Rodden or *The Gauquelin Book of American Charts* by Michel and Françoise Gauquelin.

URANUS IN THE FIRST HOUSE	
ALWORTH, Lance	GILRUTH, Robert
ANTHONY, Earl	GRAEBNER, Clark
ARNESS, James	GRAMM, Donald
BERGQUIST, Kenneth	HAYES, Bob
BERRA, Yogi	JAMES, Daniel
BREEDLOVE, Craig	KOONTZ, Elizabeth
BROWNING, John	LOVE, John
CHANDLER, Otis	LUCAS, George
CHARLES, Ezzard	Mc CARTHY, Eugene J.
DALY, James	Mc COVEY, Willy L.
DIEBOLD, John	MATHEWS, Edwin
DULLEA, Keir	MILLETT, Kate
DUVALL, Robert	MONDALE, Walter
FAIRCHILD, John	MORGAN, Joe
FARROW, Mia	PAAR, Jack
FEINSTEIN, Dianne	SLAYTON, Donald
FLEISCHER, Leon	SPRINGSTEEN, Bruce
FORD, Henry	WARMERDAM, Cornelius
FREEMAN, Orville	YARBOROUGH, Cale

URANUS IN THE SECOND HOUSE

BABASHOFF, Shirley	LAYNE, Bobby
BOMBECK, Erma	LEWIS, Henry
BOWEN, William	LILLEHEI, Clarence
BRADBURY, Ray	LOPEZ, Nancy
CHAMPION, Gower	Mc BRIDE, Patricia
CHAPLIN, Geraldine	MINNELLI, Liza
CONNALLY, John	MORTON, Roger
DAWSON, Leonard	MOYERS, Bill
DEGAETANI, Jan	NEWMAN, Paul
DONAHUE, Phil	NOVAK, Kim
DUERK, Alene	PERRINE, Valerie
FRIEDAN, Betty	PREUS, Jacob
GABRIEL, Roman	PRICE, Leontyne
GALVIN, Robert	PRYOR, Richard
GRIZZARD, George	REDFORD, Robert
HARRELSON, Ken	ROSE, Peter
HUGHES, Emmet	SIMONE, Nina
JOHNSON, Don	STEINBRENNER, George
KAHN, Madeline	STEINKRAUS, Bill
KESEY, Ken	WEINBERGER, Casper
KROL, John	WILLIAMS, Billy

URANUS IN THE THIRD HOUSE

ASH, Roy	LOGAN, Karen
BEBAN, Gary	LOVELL, James
BENTSEN, Lloyd	Mc ADOO, Bob
BRADLEY, Thomas	Mc CARTHY, Marie Therese
BRANDO, Marlon	Mc ELHENNY, Hugh E.
BROOKE, E.	MAYS, Willie
BROWER, David	MIDLER, Bette
CERNAN, Eugene	MITCHELL, Lydell
DAVIS, Glen	MUNSEL, Patrice
DELAVALLADE, Carmen	NEAL, Patricia
DILLMAN, Bradford	O'NEAL, Arthur
GOODSON, Mark	PERRY, James
GOULIAN, Mehran	RAUSCHENBERG, Robert
HAGGARD, Merle	REHNQUIST, William
HOLBROOK, Hal	RIGGS, Bobby
HORNUNG, Paul	ROBERTS, Robin
JACKSON, Henry	SEAGREN, Bob
KASTENMEIER, Robert	SHEEN, Martin
KIRKLAND, Joseph	THOMPSON, David
KLEINDIENST	WILLIAMS, Theodore
KNIEVEL, Evel	WRIGHT, Mickey
KUCINICH, Dennis	

URANUS IN THE FOURTH HOUSE

AMECHE, Don
ARFONS, Art
BANKS, Ernie
BERGLAND, Robert
BERRIGAN, Daniel
COPPOLA, Francis
CROSSFIELD, Albert
CURTIS, Ann
DAVENPORT, Willie
DEBUSSCHERE, David
EHRLICHMAN, John
ETHERINGTON, Edwin
FELKER, Clay
FODOR, Eugene
GLENN, John
GROZA, Louis
HODGSON, James
HOFFMAN, Dustin
INOUYE, Daniel
JACKSON, Maynard
JORDAN, William
KING, Billie

KITT, Eartha
LAWRENCE, Carol
LUCAS, Jerry
LUCE, Charles
Mac DONALD, Ross
MATHIAS, Robert
MEREDITH, Burgess
MOTHERWELL, Robert
MOTT, Stewart
NASH, Philleo
PAGE, Alan Cedric
RHODES, John
ROTH, Philip
SAINT, Eva Marie
SALINGER, Pierre
SAYERS, Gale
TEMPLE, Shirley
VAUGHAN, Sarah
WARFIELD, Paul
WARNECKE, John
WIDMARK, Richard
WILLSON, Zack

URANUS IN THE FIFTH HOUSE

ANDERSON, John
BERRY, Raymond
BROOKS, Gwendolyn
BUTTON, Richard
CAGE, John
CARSON, Johnny
CLIFFORD, John
CONNOLLY, Maureen
CRAIN, Jeanne
DYLAN, Bob
EASTWOOD, Clint
EVANS, Daniel
FRAZIER, Joe
GAGNE, Verne
GODDARD, James
GRAHAM, Otto
GREY, Joel
HAMILL, Dorothy
HART, Phil
HAUGE, Gabriel
HERMAN, Woody
HOLLEY, Robert
KAEL, Pauline
KIRK, Claude
KNOWLES, John
LEIGHTON, R. B.

LEWIS, Roger
LYNN, James
McCLOSKEY, Paul N.
MARIS, Roger
MURPHY, William
NABER, John
PAIGE, Janis
PETTIT, Robert
PHILLIPS, Wendell
PRESLEY, Elvis
RAUH, Joseph
REED, Willis
SCAMMON, Richard
SMEAL, Eleanor
SNYDER, Gary
STAUBACH, Roger
STEWART, Thomas
STONES, Dwight
TETLEY, Glen
THOMAS, Michael
TURNER, Ted
VOLCKER, Paul
WHITE, Edward
WILLS, Helen
YOUNG, John

URANUS IN THE SIXTH HOUSE

ANNENBERG, Walter
BACKE, John
BERGEN, Candice
BORMAN, Frank
CHANNING, Carol
CLARK, Eleanor
COHEN, Wilbur
CRONIN, Joseph
DAY, Doris
DONOVAN, Hedley
DUNLOP, John
FABRAY, Nanette
FOXX, Redd
FOYT, Anthony
GIBSON, Althea
HALDEMAN, Harry
HARRIS, Fred
HARRIS, Franco
HEARNES, W. E.

KERKORIAN, Kirk
KOCH, John
KRISTOFFERSON, Kris
Mc KISSICK, Floyd
MANN, Dick
MARTIN, Billy
MIKAN, George
MILLER, John
MILLER, Neal
MITCHELL, John
MURAYAMA, Makio
NICKLAUS, Jack
NOLAND, Kenneth
O'NEAL, Ryan
RICHARDS, Bob
SCHLESINGER, Arthur
STEINEM, Gloria
WELCH, Raquel
WYNN, Early

URANUS IN THE SEVENTH HOUSE

BLAKE, Robert
BLEGEN, Judith
BLUME, Judy
BOUTON, Jim
CHILES, Lawton
COOLEY, Denton
COSELL, Howard
CRANSTON, Alan
CSONKA, Lawrence
DINE, James
DOWNS, Hugh
EISENHOWER, John
FLEMING, Peggy
GRIFFIN, Archie
HALL, Floyd
HANSEN, Fred
HARD, Darlene
HEFNER, Hugh
HELMS, Rich
HOWARD, Frank
HUFSTEDLER, Shirley
IRVING, John
JENSEN, Arthur

JONES, Everett
KEATON, Diane
LEE, Sammy
LEMON, Meadow
LEVINE, James
LOSEY, Joseph
LYNN, Janet
McKINLEY, Chuck
MacNEIL, Cornell Hill
MALDEN, Karl
MARSHALL, Burke
MASTERS, William
PEARSON, David
RICE, Jim
ROSS, Diana
SAWHILL, John
SCHIRRA, Walter
SIX, Robert
TARKENTON, Francis
THEROUX, Paul
TOWNES, Charles
UDALL, Morris

URANUS IN THE EIGHTH HOUSE

AARON, Hank
ASHE, Arthur
BEAN, Orson
BLUE, Vida
CASPER, William
COLBY, William
DEMPSEY, Tom
DRYSDALE, Don
GARDNER, Ava
GARDNER, John
GRONOUSKI, John
HARTZOG, George
HILLER, Stanley
LEDERBERG, Joshua
LILLY, Bob
LITTLER, Eugene
MATSON, Oliver

MORIARTY, Michael
OKUN, Arthur
RAY, Dixy
SAVITT, Richard
SCHMIDT, Mike
SEAVER, Tom
SHAPP, Milton
SHOEMAKER, Eugene
SIMON, William
STEVENS, George
STOKES, Carl
TALESE, Gay
TYUS, Wyomia
WELLS, Mary
WEYERHAEUSER, George
WICKER, Tom
YOUNG, Whitney

URANUS IN THE NINTH HOUSE

ALPERT, Herb
ANDERSON, Jack
BALL, William
BLANCHARD, Felix
BOONE, Richard
BRADLEY, Bill
BRUBECK, David
BUDGE, John
CAULFIELD, Joan
CAUTHEN, Steve
CHADWICK, Florence
CHAMBERLAIN, Owen
CHILD, Julia
COWENS, David
EAGLETON, Thomas
ELDER, Lee
FIDRYCH, Mark
FIELD, Sally
FRANCIS, Sam
GABLE, Dan
JURGENSEN, Christine
KRAMER, John
LEWIS, David
LEWIS, Jerry

LYNDE, Paul
MATSON, Randi
MAYO, Robert
MOORE, George
MUNSON, Thurman
NELSON, Willie
OATES, J. C.
O'BRIEN, Parry
OTTO, James
OWEN, Nancy
PARSEGHIAN, Ara
PECKINPAH, Samuel
PLUNKETT, Jim
POSTON, Tom
REVELLE, Roger
RICHARDSON, Bobby
RUSH, David
RYUN, James
SHORE, Dinah
WELLES, Orson
WHITE, Byron
WHITE, Jojo
WINTERS, Jonathan

URANUS IN THE TENTH HOUSE

ALI, Muhammed
ALVAREZ, Luis
ASHLEY, Elisabeth
BAILAR, Benjamin
BARRY, Richard
BROWN, Jerry
BURKE, Yvonne
BUTKUS, Dick
COLLINS, Judy
CONIGLIARO, Tony
DILLER, Phyllis
DUNCAN, Charles
FOREMAN, George
GARLAND JUDY
GIFFORD, Frank
GLASER, D. A.
GOOD, Robert
GREENE, Joe
GRIFFIN, Merv
JONES, Deacon

KRAMER, Gerald
LAMB, William
LOWENSTEIN, Allard
Mc WHINNEY, Madeline
MAUCH, Gene
NITSCHKE, Raymond
O'HAIR, Madalyn
PARKS, Robert
PEROT, Henry
RAINEY, Froelich
RALSTON, R.
ROZELLE, P.
SCHOLLANDER, Donald
SHORT, Bobby
SUMMER, Donna
TAYLOR, James
TRABERT, Marion
WHITWORTH, Kathy
WILHELM, James
WYETH, James

URANUS IN THE ELEVENTH HOUSE

ARMSTRONG, Neil	MARTIN, Dean
ASKEW, R.	MAY, Rollo
BYRNE, Brendan	MOYNIHAN, Daniel
DAVIS, Miles	OLSEN, Merlin
DELLUMS, Ronald	SCHLAFLY, Phillis
FARREL, Suzanne	SEITZ, Frederick
GOLD, Herbert	SHAPIRO, Irving
GRAHAM, Billy	SIMPSON, Orenthal J.
GREGORY, Cynthia	SPITZ, Mark
HAACK, Robert	STARR, Bryan
HARRIS, Patricia	TALBERT, Bill
HAYES, Elvin	TARR, Curtis
KEMP, Jack	VALENTI, Jack
KIDD, Billy	WAGNER, A. J.
LAMBERT, Jack	ZUMWALT, Elmo

URANUS IN THE TWELFTH HOUSE

ALIOTO, Joseph	MAILER, Norman
BAKER, Howard	MORRISON, Toni
BLACK, Karen	MURRAY, Don
BYRD, Robert	PAINE, Thomas
COUNSILMAN, James	PATTERSON, Floyd
DELOREAN, John	PEPPLER, Mary
DENVER, John	RATHER, Dan
FINGERS, Rollie	ROBERTS, Ken
FOSBURY, Richard	ROGERS, Roy
FREEMAN, David	SCALI, John
GILLIGAN, John	SCHROEDER, Frederick
GONZALES, Richard	SCOTT, David
GOODMAN, Julian	SPITZER, Lyman
GRIER, Rosey	STEWART, William
HADDON, William	TAFT, Robert
HAVLICEK, John	TRAVOLTA, John
HILLS, Carla	TREVINO, Lee
JANOV, Arthur	TROWBRIDGE, Alexandre
KILMER, Bill	VERDON, Gwen
LONDON, Julie	WALTON, Bill
Mc CAMBRIDGE, Mecedes	WARWICK, Dionne
MAHAN, Larry	ZIEGLER, Ronald

URANUS IN ARIES

0♈26	KITT, Eartha	15♈17 R	ARMSTRONG, Neil
1♈16	McCLOSKEY, Paul	15♈23 R	LITTLER, Eugene
2♈12	VOLCKER, Paul	15♈57	O'BRIEN, Parry
2♈35 R	GIBSON, Althea	16♈16	TALESE, Gay
2♈50	BORMAN, Frank	16♈35	RATHER, Dan
3♈17	STOKES, Carl	16♈39	BALL, William
3♈25 R	LOVELL, James	17♈01	WILLSON, Zack
3♈33	OKUN, Arthur	17♈03 R	MAYS, Willie
3♈37	McELHENNY, Hugh	17♈18 R	MATHEWS, Edwin
3♈49	GRIZZARD, George	18♈59 R	CASPER, William
3♈59	LOWENSTEIN, Allard	19♈01	STEVENS, George
5♈05 R	TEMPLE, Shirley	19♈29	GREY, Joel
5♈19	SHOEMAKER, E.	19♈34	PETTIT, Robert
5♈49 R	GONZALES, Richard	20♈21 R	NOVAK, Kim
6♈04	ASKEW, R.	20♈40 R	SIMONE, Nina
6♈09 R	MARTIN, Billy	20♈58 R	BERRY, Raymond
6♈29 R	WELLS, Mary	21♈19 R	LEWIS, Henry
6♈33 R	STEWART, Thomas	21♈45	BURKE, Yvonne
7♈21	FLEISCHER, Leon	21♈55	ROTH, Philip
7♈21	BERGLAND, Robert	22♈18 R	SCOTT, David
7♈21 R	BEAN, Orson	22♈49 R	LAWRENCE, Carol
7♈28	TROWBRIDGE, A.	23♈10	BACKE, John
7♈32	GOULIAN, Mehran	23♈18	GRIER, Rosey
9♈28	DUVALL, Robert	23♈28	HILLS, Carla
10♈29 R	EAGLETON, Thomas	23♈29	STARR, Bryan
11♈18	MURRAY, Don	23♈59	AARON, Hank
11♈23	BUTTON, Richard	24♈03 R	LEMON, Meadow
11♈27	CHILES, Lawton	24♈19 R	NELSON, Willie
11♈55 R	MATHIAS, Robert	25♈28 R	CERNAN, Eugene
11♈60	WHITE, Edward	25♈30	BROWNING, John
12♈02	HARRIS, Fred	25♈55	BOWEN, William
12♈03	DILLMAN, Bradford	25♈57 R	AMECHE, Don
12♈07	BANKS, Ernie	26♈05 R	STEINEM, Gloria
12♈52	MORRISON, Toni	26♈35	BLAKE, Robert
12♈60 R	MATSON, Oliver	26♈42 R	FEINSTEIN, Dianne
13♈22	SNYDER, Gary	27♈13 R	DEGAETANI, Jan
13♈38	DELAVALLADE, C.	27♈30	PATTERSON, Floyd
13♈53	YOUNG, John	27♈30 R	PRESLEY, Elvis
14♈23	EASTWOOD, Clint	27♈32	FOYT, Anthony
15♈07	GIFFORD, Frank	27♈35	BAILAR, Benjamin
15♈07 R	TRABERT, Marion	27♈36 R	PEARSON, David
15♈08	PEROT, Henry	28♈09	WRIGHT, Mickey
15♈16	STEINBRENNER, G.	29♈59 R	MOYERS, Bill
		0♉06 R	JONES, Everett

URANUS IN TAURUS

0♉12	ALPERT, Herb	15♉30	YARBOROUGH, Cale
0♉19	MANN, Dick	15♉57 R	WILLIAMS, Billy
0♉47	CONNOLLY, Maureen	16♉01	OATES, J. C.
0♉52 R	MILLETT, Kate	16♉02	COPPOLA, Francis
0♉59	MARIS, Roger	16♉41 R	KNIEVEL, Evel
1♉14 R	ELDER, Lee	17♉24 R	COLLINS, Judy
1♉21	JURGENSEN, Christine	17♉60	NICKLAUS, Jack
1♉33 R	HARD, Darlene	18♉00	TARKENTON, Francis
1♉38 R	KRAMER, Gerald	18♉01 R	ZIEGLER, Ronald
1♉41	HORNUNG, Paul	19♉12	TREVINO, Lee
1♉43 R	DONAHUE, Phil	19♉37 R	LUCAS, Jerry
2♉25	DELLUMS, Ronald	20♉05	HAVLICEK, John
4♉18	DINE, James	20♉38 R	BLACK, Karen
4♉28	DAWSON, Leonard	21♉31	LILLY, Bob
4♉58	KESEY, Ken	21♉36 R	WHITWORTH, Kathy
5♉10	KEMP, Jack	21♉38	SMEAL, Eleanor
5♉29 R	RICHARDSON, Bobby	21♉56 R	KILMER, Bill
5♉43	NITSCHKE, Raymond	21♉58 R	ASHLEY, Elisabeth
6♉36	O'NEAL, Arthur	22♉22 R	JOHNSON, Don
7♉24 R	DULLEA, Keir	22♉25	McKINLEY, Chuck
7♉30	BREEDLOVE, Craig	22♉35	HANSEN, Fred
7♉36 R	PERRY, James	23♉06	WARWICK, Dionne
8♉03 R	SAWHILL, John	23♉31 R	PRYOR, Richard
8♉13	HAGGARD, Merle	23♉54 R	MORIARTY, Michael
8♉29 R	KRISTOFFERSON, Kris	24♉09 R	THEROUX, Paul
9♉23	DRYSDALE, Don	24♉41	O'NEAL, Ryan
9♉34	HOWARD, Frank	25♉04	BLEGEN, Judith
9♉35 R	REDFORD, Robert	25♉20 R	DEBUSSCHERE, David
9♉44 R	McCOVEY, Willy	25♉49	SHEEN, Martin
9♉47	OTTO, James	25♉49	ALWORTH, Lance
10♉00	BLUME, Judy	25♉52 R	GABRIEL, Roman
10♉31	MOTT, Stewart	26♉04	OLSEN, Merlin
11♉23	JACKSON, Maynard	26♉08 R	WELCH, Raquel
12♉10 R	BROWN, Jerry	26♉21	STAUBACH, Roger
13♉16 R	ANTHONY, Earl	26♉28 R	ALI, Muhammed
13♉39 R	HOFFMAN, Dustin	26♉38 R	DYLAN, Bob
14♉36	JONES, Deacon	26♉40	IRVING, John
14♉43 R	BOUTON, Jim	28♉12 R	ROSE, Peter
15♉22	TURNER, Ted		

URANUS IN GEMINI

0♊21 R	HARRELSON, Ken	13♊35	MINNELLI, Liza
1♊33	HAYES, Bob	14♊17	KEATON, Diane
1♊58	BUTKUS, Dick	15♊21	SCHOLLANDER, D.
2♊17	KIDD, Billy	15♊40	MIDLER, Bette
2♊22	REED, Willis	15♊48 R	BERGEN, Candice
2♊26	WARFIELD, Paul	16♊14 R	HAYES, Elvin
3♊17	OWEN, Nancy	16♊34 R	PAGE, Alan Cedric
3♊46	RALSTON, R.	16♊51	FARREL, Suzanne
4♊16 R	SAYERS, Gale	17♊12	TYUS, Wyomia
4♊26	KAHN, Madeline	17♊47 R	FOSBURY, Richard
4♊28	McBRIDE, Patricia	18♊32 R	DEMPSEY, Tom
5♊07	FRAZIER, Joe	19♊09	WYETH, James
5♊26	DAVENPORT, W.	19♊13	CSONKA, Lawrence
5♊35	DENVER, John	19♊16	GREGORY, Cynthia
5♊39	ROSS, Diana	19♊26	RYUN, James
5♊42	BARRY, Richard	19♊26	MILLER, John
6♊16	LEVINE, James	20♊39 R	BEBAN, Gary
7♊07 R	ASHE, Arthur	20♊53	WHITE, Jojo
7♊07	KING, Billie	21♊09 R	FIELD, Sally
7♊09 R	MAHAN, Larry	21♊19	FINGERS, Rollie
7♊50	GRAEBNER, Clark	21♊34 R	MUNSON, Thurman
7♊54	BRADLEY, Bill	21♊38	SEAGREN, Bob
7♊60	LUCAS, George	21♊45 R	KUCINICH, Dennis
8♊48	PERRINE, Valerie	21♊48	GREENE, Joe
8♊50 R	MORGAN, Joe	21♊53	CLIFFORD, John
9♊07	FARROW, Mia	22♊11 R	TAYLOR, James
9♊15	MATSON, Randi	23♊26 R	SIMPSON, O.J.
9♊44 R	CONIGLIARO, Tony	24♊38	PLUNKETT, Jim
10♊21 R	THOMAS, Michael	27♊39 R	FOREMAN, George
11♊46	SEAVER, Tom	28♊03 R	SUMMER, Donna
12♊11 R	CHAPLIN, Geraldine	28♊36 R	FLEMING, Peggy
12♊46 R	PEPPLER, MARY	29♊22 R	MITCHELL, Lydell
13♊08	JORDAN, William		

URANUS IN CANCER

0♋01	LOGAN, Karen	13♋43	McADOO, Bob
0♋25 R	GABLE, Dan	14♋21	LAMBERT, Jack
0♋25 R	COWENS, David	14♋32	LYNN, Janet
0♋56 R	HARRIS, Franco	14♋32 R	RICE, Jim
0♋57	FODOR, Eugene	18♋28 R	WALTON, Bill
1♋15 R	SPITZ, Mark	19♋35	TRAVOLTA, John
2♋49 R	BLUE, Vida	22♋29 R	STONES, Dwight
4♋53	SPRINGSTEEN, Bruce	23♋10	THOMPSON, D.
4♋56 R	SCHMIDT, Mike	25♋06	FIDRYCH, Mark
12♋03 R	ROBERTS, Ken	25♋30 R	GRIFFIN, Archie

URANUS IN LEO

0♌19	NABER, John	5♌43 R	LOPEZ, Nancy
2♌43	HAMILL, Dorothy	16♌57	CAUTHEN, Steve
4♌39 R	BABASHOFF, Shirley		

No placements in Virgo, Libra, Scorpio or Sagittarius

URANUS IN CAPRICORN

0♑33 R	WILLS, Helen	21♑04	MAY, Rollo
4♑50	CRONIN, Joseph	21♑29 R	RUSH, David
10♑07	MEREDITH, Burgess	21♑32	KROL, John
11♑04 R	SIX, Robert	24♑33 R	RAUH, Joseph
11♑21 R	RAINEY, Froelich	25♑48 R	ROGERS, Roy
11♑22	MURPHY, William	28♑00	SEITZ, Frederick
16♑17	ANNENBERG, Walter	28♑45	ALVAREZ, Luis
17♑25 R	LOSEY, Joseph	28♑54	LEWIS, Roger
17♑27 R	NASH, Philleo	28♑56	WAGNER, A. J.
17♑44	KOCH, John	29♑31	GARDNER, John
18♑14	MILLER, Neal	29♑59	CAGE, John
20♑05 R	REVELLE, Roger		

URANUS IN AQUARIUS

0♒18 R	BERGQUIST, Kenneth	12♒28	HODGSON, James
0♒37	CHILD, Julia	12♒50 R	MacDONALD, Ross
0♒49	MURAYAMA, Makio	13♒29 R	MASTERS, William
1♒06	HART, Phil	14♒05	TOWNES, Charles
2♒22	BROWER, David	14♒30	SCAMMON, Richard
2♒35	SHAPP, Milton	15♒15 R	WARMERDAM, C.
2♒42	McCARTHY, Marie T.	15♒28	BUDGE, John
3♒14	JACKSON, Henry	15♒35	WELLES, Orson
3♒37 R	GILRUTH, Robert	16♒05	ALIOTO, Joseph
4♒12 R	MITCHELL, J.	16♒17	LOVE, John
6♒14 R	LAMB, William	16♒20 R	RHODES, John
6♒27	CLARK, Eleanor	17♒03	SHORE, Dinah
6♒48	HELMS, Rich	17♒49	MAYO, Robert
7♒13 R	COHEN, Wilbur	18♒27 R	McCARTHY, Eugene
7♒32	HERMAN, Woody	18♒42 R	SHAPIRO, Irving
8♒01 R	MORTON, Roger	18♒43	HALL, Floyd
8♒28	RAY, Dixy	19♒53 R	SCHLESINGER, Arthur
9♒34 R	WIDMARK, Richard	19♒59	BYRD, Robert
9♒44	HAUGE, Gabriel	20♒05	HAACK, Robert
10♒42 R	DUNLOP, John	20♒42	TAFT, Robert
10♒58	SPITZER, Ly	20♒45 R	CONNALLY, John
11♒10 R	GOODSON, Mark	20♒56	FORD, Henry
11♒11 R	CRANSTON, Alan	21♒10	LUCE, Charles
11♒13	MOTHERWELL, Robert	21♒15 R	BRADLEY, Thomas
11♒36 R	DONOVAN, Hedley	21♒35	WEINBERGER, Casper

URANUS IN AQUARIUS

22≈50 R DILLER, Phyllis	25≈05 R TALBERT, Bill
23≈10 LEWIS, David	25≈18 WILLIAMS, Theodore
23≈25 PAAR, Jack	25≈30 McCAMBRIDGE, M.
23≈33 BOONE, Richard	27≈11 SCALI, John
23≈34 MARTIN, Dean	27≈29 FREEMAN, Orville
23≈34 R BROOKS, Gwendolyn	27≈47 MALDEN, Karl
23≈40 R WHITE, Byron	27≈50 R GRONOUSKI, John
23≈41 R KERKORIAN, Kirk	27≈50 R BROOKE, E.
23≈48 R GRAHAM, Billy	28≈09 WARNECKE, John
23≈49 CHADWICK, Florence	29≈02 LEIGHTON, R. B.
23≈51 DALY, James	29≈08 R COLBY, William
23≈51 LILLEHEI, Clarence	29≈14 WYNN, Early
23≈52 ASH, Roy	29≈18 PREUS, Jacob
24≈24 R RIGGS, Bobby	

URANUS IN PISCES

0♓33 O'HAIR, Madalyn	6♓42 R PHILLIPS, Wendell
1♓05 R JAMES, Daniel	7♓05 GILLIGAN, John
1♓38 R KAEL, Pauline	7♓26 R VALENTI, Jack
1♓42 KOONTZ, Elizabeth	7♓53 HOLLEY, Robert
1♓44 MOORE, George	8♓47 R KRAMER, John
1♓51 R FABRAY, Nanette	8♓49 R YOUNG, Whitney
1♓54 R ZUMWALT, Elmo	8♓51 ANDERSON, John
2♓02 BRUBECK, David	9♓08 R SCHROEDER, F.
2♓36 R HUGHES, Emmet	9♓08 BERRIGAN, Daniel
2♓41 COUNSILMAN, James	9♓11 R GLENN, John
3♓05 HARTZOG, George	9♓22 R STEWART, William
3♓16 R FREEMAN, David	9♓25 CHARLES, Ezzard
3♓30 R COSELL, Howard	9♓37 CHAMPION, Gower
3♓41 R DUERK, Alene	9♓50 FOXX, Redd
3♓52 R BRADBURY, Ray	10♓07 McKISSICK, Floyd
3♓53 R COOLEY, Denton	10♓12 GARDNER, Ava
4♓18 CHANNING, Carol	10♓12 McWHINNEY, Madeline
4♓29 FRIEDAN, Betty	10♓14 KIRKLAND, Joseph
4♓40 LEE, Sammy	10♓19 R ANDERSON, Jack
4♓52 R BENTSEN, Lloyd	10♓20 GALVIN, Robert
5♓04 R DOWNS, Hugh	10♓37 MARSHALL, Burke
5♓19 CHAMBERLAIN, Owen	10♓52 MacNEIL, Cornell H.
5♓44 R PAINE, Thomas	11♓09 PAIGE, Janis
5♓54 R GRAHAM, Otto	11♓21 PARKS, Robert
6♓03 POSTON, Tom	11♓28 R DAY, Doris
6♓28 R CROSSFIELD, Albert	11♓45 MAILER, Norman

URANUS IN PISCES (CONTINUED)

12♓41 R GOODMAN, Julian	22♓08 R KIRK, Claude
12♓49 R EISENHOWER, John	22♓09 CARSON, Johnny
13♓04 WILHELM, James	22♓11 R RAUSCHENBERG, R.
13♓17 GOOD, Robert	22♓22 R EVANS, Daniel
13♓29 R CAULFIELD, Joan	22♓30 R STEINKRAUS, Bill
13♓34 GARLAND, Judy	22♓35 R NEAL, Patricia
13♓36 R UDALL, M.	22♓52 R FELKER, Clay
13♓59 R SCHIRRA, Walter	23♓11 R TETLEY, Glen
15♓09 R KASTENMEIER, Robert	23♓11 ARFONS, Art
15♓12 R GROZA, Louis	24♓06 RICHARDS, Bob
16♓11 R JENSEN, Arthur	24♓23 R HUFSTEDLER, Shirley
16♓12 GODDARD, James	24♓26 GAGNE, Verne
16♓50 KLEINDIENST	24♓31 BERRA, Yogi
17♓07 SLAYTON, Donald	24♓35 R ROZELLE, Pete
17♓08 R PARSEGHIAN, Ara	24♓37 MUNSEL, P.
17♓10 R HEARNES, W. E.	24♓53 R CURTIS, Ann
17♓15 ARNESS, James	24♓54 R LEDERBERG, Joshua
17♓33 FRANCIS, Sam	24♓58 R CRAIN, Jeanne
17♓34 R GOLD, Herbert	25♓22 SALINGER, Pierre
17♓39 HILLER, Stanley	25♓25 R LEWIS, Jerry
17♓41 BLANCHARD, Felix	25♓28 R GRIFFIN, Merv
17♓56 ETHERINGTON, Edwin	25♓36 R LAYNE, Bobby
17♓58 R DAVIS, Glen	26♓08 HALDEMAN, Harry
18♓17 R DELOREAN, John	26♓46 HEFNER, Hugh
18♓32 VERDON, Gwen	27♓05 R ROBERTS, Robin
18♓37 VAUGHAN, Sarah	27♓13 R LONDON, Julie
18♓47 REHNQUIST, William	27♓18 R PRICE, Leontine
18♓51 BYRNE, Brendan	27♓25 R GLASER, D. A.
19♓00 BRANDO, Marlon	27♓51 R BOMBECK, Erma
19♓04 R NEWMAN, Paul	27♓54 DUNCAN, Charles
19♓16 TARR, Curtis	28♓07 GRAMM, Donald
19♓21 R NOLAND, Kenneth	28♓13 R LYNN, James
19♓23 SHORT, Bobby	28♓29 SAVITT, Richard
19♓43 INOUYE, Daniel	28♓34 R FAIRCHILD, John
20♓09 R HOLBROOK, Hal	28♓44 KNOWLES, John
20♓22 R JANOV, Arthur	28♓45 HADDON, William
20♓24 PECKINPAH, Samuel	28♓47 R DAVIS, Miles
20♓34 R SCHLAFLY, Phillis	29♓08 MOYNIHAN, D.
21♓14 HARRIS, Patricia	29♓09 DIEBOLD, John
21♓29 SAINT, Eva Marie	29♓15 LYNDE, Paul
21♓29 R MIKAN, George	29♓20 WICKER, Tom
21♓37 MAUCH, Gene	29♓26 R WEYERHAEUSER, G.
21♓39 BAKER, Howard	29♓33 R SIMON, William
21♓43 WINTERS, Jonathan	29♓36 CHANDLER, Otis
21♓58 EHRLICHMAN, John	29♓48 MONDALE, Walter

SUN/URANUS

CONJUNCTION ☌ 4° orb

0°03'	LEWIS, Jerry	1°59'	CHILES, Lawton
0°30'	CLIFFORD, John	1°59'	LAMBERT, Jack
0°58'	LOVELL, James	2°30'	GREY, Joel
1°08'	MOORE, George	2°48'	THOMPSON, D.
1°12'	GOLD, Herbert	2°51'	ZIEGLER, Ronald
1°21'	HAMILL, Dorothy	3°16'	BAILAR, Benjamin

SEXTILE ✳ 3° orb

0°11'	RALSTON, R.	2°18'	WELLS, Mary
0°43'	ROSS, Diana	2°22'	HODGSON, James
0°49'	MUNSEL, P.	2°23'	DELOREAN, John
1°15'	BERGQUIST, Kenneth	2°26'	GOODMAN, Julian
1°28'	PAGE, Alan Cedric	2°33'	BANKS, Ernie
1°45'	TALESE, Gay	2°37'	HELMS, Rich
1°49'	HUGHES, Emmet	2°48'	FEINSTEIN, Dianne
2°15'	BARRY, Richard	2°53'	BOUTON, Jim

SQUARE □ 3° orb

0°04'	TAYLOR, James	1°54'	PERRINE, Valerie
0°40'	WELLES, Orson	1°59'	SALINGER, Pierre
0°45'	SCHMIDT, Mike	2°02'	BLANCHARD, Felix
1°14'	KRAMER, Gerald	2°18'	FOYT, Anthony
1°22'	LAYNE, Bobby	2°22'	FOSBURY, Richard
1°38'	LYNN, Janet	2°23'	HOFFMAN, Dustin
1°44'	GRIER, Rosey	2°45'	WICKER, Tom
1°48'	GOULIAN, Mehran	2°56'	STEINBRENNER, George

TRINE △ 3° orb

0°41'	PETTIT, Robert	2°09'	PEARSON, David
0°49'	ALI, Muhammed	2°23'	SEAGREN, Bob
0°50'	HORNUNG, Paul	2°24'	BROOKS, Gwendolyn
0°58'	CHANDLER, Otis	2°26'	NITSCHKE, Raymond
1°20'	BAKER, Howard	2°31'	WINTERS, Jonathan
1°27'	COWENS, David	2°35'	ASH, Roy
1°30'	SCHLESINGER, Arthur	2°48'	MURRAY, Don
1°34'	GABLE, Dan	2°50'	DONAHUE, Phil
1°42'	KAHN, Madeline	2°51'	MARTIN, Dean
1°45'	JURGENSEN, Christine	2°53'	RICE, Jim
1°53'	FABRAY, Nanette	0°30'	PERRY, James

SUN/URANUS (CONTINUED)

OPPOSITION ☍ 4° orb

0°46'	NABER, John	2°30'	ROBERTS, Ken
0°60'	GLASER, D. A.	3°21'	WARFIELD, Paul
1°60'	LEWIS, Henry	3°38'	SHORT, Bobby
2°06'	DUVALL, Robert	3°48'	WEINBERGER, Casper
2°18'	MATHEWS, Edwin		

MOON/URANUS

CONJUNCTION ☌ 4° orb

0°00'	LOWENSTEIN, Allard	2°03'	BERRY, Raymond
0°13'	SAVITT, Richard	2°05'	BLEGEN, Judith
0°19'	BARRY, Richard	3°03'	BROWER, David
1°23'	GILRUTH, Robert	3°14'	WELLES, Orson
2°01'	RICHARDSON, Bobby	3°45'	MORGAN, Joe

SEXTILE ⚹ 3° orb

0°11'	FABRAY, Nanette	1°47'	WHITWORTH, Kathy
0°52'	CHILES, Lawton	1°47'	KIDD, Billy
1°20'	FRIEDAN, Betty	1°49'	O'NEAL, Arthur
1°25'	CHAMBERLAIN, Owen	2°08'	WARNECKE, John
1°33'	PAGE, Alan Cedric	2°16'	POSTON, Tom
1°43'	SLAYTON, Donald	2°44'	MURRAY, Don
1°43'	MIKAN, George		

SQUARE □ 3° orb

0°15'	CRANSTON, Alan	1°56'	LAMB, William
0°45'	BAILAR, Benjamin	1°57'	LEWIS, Roger
0°51'	HEARNES, W. E.	2°04'	GRAMM, Donald
1°02'	WEYERHAEUSER, G.	2°10'	SIMPSON, O.J.
1°17'	OKUN, Arthur	2°27'	RUSH, David
1°21'	FODOR, Eugene	2°32'	KRAMER, Gerald
1°22'	MEREDITH, Burgess	2°38'	PEROT, Henry
1°25'	CONNALLY, John	2°38'	SIMONE, Nina
1°42'	KUCINICH, Dennis	2°53'	KNIEVEL, Evel
1°53'	MAYS, Willie		

MOON/URANUS (CONTINUED)

TRINE △ 3° orb

0°11'	Mc CARTHY, Marie T.	1°50'	COLBY, William
0°53'	DALY, James	1°54'	RALSTON, R.
1°05'	VALENTI, Jack	1°55'	RHODES, John
1°15'	HERMAN, Woody	2°08'	WILLIAMS, Billy
1°16'	FREEMAN, David	2°10'	GARDNER, John
1°17'	WYETH, James	2°13'	ROBERTS, Robin
1°23'	LYNDE, Paul	2°16'	REHNQUIST, William
1°39'	GRAEBNER, Clark	2°49'	Mc BRIDE, Patricia
1°44'	LILLEHEI, Clarence	2°58'	REVELLE, Roger
1°48'	HILLER, Stanley	2°59'	MURPHY, William

OPPOSITION ☍ 4° orb

0°24'	DELAVALLADE, Carmen	1°35'	GROZA, Louis
0°57'	SAYERS, Gale	2°11'	SEAVER, Tom
1°07'	BEAN, Orson	3°18'	REED, Willis
1°09'	WICKER, Tom	3°27'	BERGLAND, Robert

MERCURY/URANUS

CONJUNCTION ☌ 4° orb

0°25'	GREY, Joel	1°53'	GILLIGAN, John
0°51'	SAYERS, Gale	2°02'	LYNN, James
0°59'	SAVITT, Richard	2°09'	ROZELLE, P.
1°06'	RAUH, Joseph	2°46'	JOHNSON, Don
1°08'	RIGGS, Bobby	3°14'	GRAMM, Donald
1°17'	OWEN, Nancy	3°24'	LEVINE, James
1°24'	FAIRCHILD, John		

SEXTILE ✶ 3° orb

0°12'	GABRIEL, Roman	2°11'	DINE, James
0°49'	WYNN, Early	2°20'	GROZA, Louis
0°52'	ALWORTH, Lance	2°20'	KASTENMEIER, Robert
0°54'	SHEEN, Martin	2°24'	MORRISON, Toni
1°08'	PREUS, Jacob	2°26'	WRIGHT, Mickey
1°14'	PEROT, Henry	2°31'	GARDNER, Ava
1°14'	HAVLICEK, John	2°35'	RYUN, James
1°31'	SCHOLLANDER, Donald	2°43'	LUCAS, Jerry
2°03'	HELMS, Rich	2°58'	FREEMAN, Orville

MERCURY/URANUS (CONTINUED)

SQUARE □ 3° orb

0°12'	HERMAN, Woody	1°45'	KOONTZ, Elizabeth
0°13'	KROL, John	2°05'	BEAN, Orson
0°15'	HARD, Darlene	2°08'	JORDAN, William
0°32'	KERKORIAN, Kirk	2°11'	SMEAL, Eleanor
0°35'	BUTTON, Richard	2°28'	BERGLAND, Robert
1°04'	STAUBACH, Roger	2°36'	BANKS, Ernie
1°04'	STEWART, William	2°39'	LILLY, Bob
1°23'	WHITE, Byron	2°42'	KIRK, Claude
1°45'	FOSBURY, Richard	2°58'	TARKENTON, Francis

TRINE △ 3° orb

0°15'	DONOVAN, Hedley	1°58'	GLENN, John
0°41'	RICHARDSON, Bobby	2°11'	SCHROEDER, Frederick
1°05'	HALDEMAN, Harry	2°17'	FARROW, Mia
1°36'	MURRAY, Don	2°22'	LAWRENCE, Carol
1°50'	HARRIS, Franco	2°28'	STONES, Dwight
1°51'	ALVAREZ, Luis	2°47'	CROSSFIELD, Albert
1°53'	PERRINE, Valerie	2°50'	PEARSON, David
1°57'	HOFFMAN, Dustin	2°60'	TAYLOR, James

OPPOSITION ☍ 4° orb

0°11'	SCHLAFLY, Phillis	2°40'	REHNQUIST, William
0°33'	KING, Billie	3°09'	GLASER, D. A.
1°08'	MATHEWS, Edwin	3°19'	LEIGHTON, R. B.
1°10'	SEAVER, Tom	3°42'	CLARK, Eleanor
1°29'	HAYES, Elvin	3°43'	EAGLETON, Thomas
1°41'	MAHAN, Larry	3°47'	BURKE, Yvonne
1°59'	WARFIELD, Paul	3°50'	PRYOR, Richard
2°09'	CSONKA, Lawrence	3°52'	ASKEW, R.
2°24'	INOUYE, Daniel		

VENUS/URANUS

CONJUNCTION ☌ 4° orb

0°25'	EHRLICHMAN, John	2°42'	RAUH, Joseph
0°46'	WILLSON, Zack	2°45'	CONNALLY, John
0°58'	BRADLEY, Thomas	3°13'	HART, Phil
1°31'	FLEMING, Peggy	3°34'	LOGAN, Karen
1°48'	GRAMM, Donald	3°47'	MAYS, Willie
1°49'	TAFT, Robert	3°48'	LYNN, James
2°01'	BREEDLOVE, Craig	3°52'	BERGEN, Candice

VENUS/URANUS (CONTINUED)

SEXTILE ✶ 3° orb

0°02'	LILLY, Bob	1°23'	BERRA, Yogi
0°11'	BLUE, Vida	1°46'	FARROW, Mia
0°18'	BUTTON, Richard	2°12'	SHORE, Dinah
0°28'	STARR, Bryan	2°22'	HILLS, Carla
1°01'	WELCH, Raquel	2°27'	WYETH, James
1°07'	TARKENTON, Francis	2°30'	VAUGHAN, Sarah
1°20'	GILLIGAN, John	2°55'	SNYDER, Gary

SQUARE □ 3° orb

0°07'	MORRISON, Toni	1°52'	PRESLEY, Elvis
1°06'	COHEN, Wilbur	2°02'	DRYSDALE, Don
1°28'	DELOREAN, John	2°08'	FABRAY, Nanette
1°34'	COLBY, William	2°41'	PATTERSON, Floyd
1°35'	GLENN, John	2°58'	KROL, John

TRINE △ 3° orb

0°00'	FLEISCHER, Leon	1°41'	BERGLAND, Robert
0°10'	BLANCHARD, Felix	1°48'	PAIGE, Janis
0°18'	KEMP, Jack	2°06'	THOMAS, Michael
0°23'	REDFORD, Robert	2°08'	BEAN, Orson
0°29'	HEARNES, W. E.	2°09'	DALY, James
0°35'	CAGE, John	2°11'	MONDALE, Walter
0°36'	TROWBRIDGE, Alexandre	2°14'	LILLEHEI, Clarence
1°26'	MARIS, Roger	2°17'	LAMB, William
1°30'	ROGERS, Roy	2°38'	JACKSON, Henry
1°37'	OTTO, James	2°52'	KERKORIAN, Kirk

OPPOSITION ☍ 4° orb

0°09'	WILHELM, James	2°51'	ASKEW, R.
0°27'	FOREMAN, George	2°59'	WARWICK, Dionne
0°29'	MITCHELL, John	3°29'	ROBERTS, Robin
1°27'	CROSSFIELD, Albert	3°53'	TALBERT, Bill
2°41'	SPITZER, Lyman	0°04'	FRAZIER, Joe

MARS/URANUS

CONJUNCTION ♂ 4° orb

0°09'	WELLS, Mary	1°38'	WILLS, Helen
0°09'	DENVER, John	1°40'	BAILAR, Benjamin
0°16'	LYNDE, Paul	2°19'	SNYDER, Gary
0°19'	GARDNER, Ava	2°49'	PERRINE, Valerie
0°19'	IRVING, John	3°07'	WICKER, Tom
0°47'	FARREL, Suzanne	3°27'	JACKSON, Maynard
0°56'	BLUE, Vida	3°53'	DIEBOLD, John

SEXTILE ✳ 3° orb

0°05'	LEWIS, Jerry	1°51'	BOMBECK, Erma
0°10'	MARSHALL, Burke	1°51'	OTTO, James
0°14'	TALESE, Gay	2°07'	ELDER, Lee
0°24'	ARMSTRONG, Neil	2°13'	LYNN, Janet
0°25'	WELLES, Orson	2°19'	TAYLOR, James
0°57'	HERMAN, Woody	2°27'	NOLAND, Kenneth
1°17'	BRANDO, Marlon	2°48'	BYRNE, Brendan
1°47'	O'HAIR, Madalyn		

SQUARE □ 3° orb

0°02'	KIRKLAND, Joseph	1°22'	WILHELM, James
0°08'	STEWART, William	1°48'	FREEMAN, David
0°09'	Mc WHINNEY, Madeline	1°51'	BUDGE, John
0°15'	CHAPLIN, Geraldine	2°06'	LAWRENCE, Carol
0°20'	EISENHOWER, John	2°30'	YOUNG, John
0°33'	KERKORIAN, Kirk	2°31'	WHITE, Byron
0°45'	STONES, Dwight	2°38'	BABASHOFF, Shirley
0°49'	Mc KISSICK, Floyd	2°40'	TETLEY, Glen
1°07'	KROL, John	2°41'	ARFONS, Art

TRINE △ 3° orb

0°12'	DILLER, Phyllis	2°01'	HARRIS, Fred
0°30'	MATHIAS, Robert	2°14'	ASHLEY, Elisabeth
0°45'	FRANCIS, Sam	2°26'	JORDAN, William
1°18'	TOWNES, Charles	2°29'	JAMES, Daniel
1°38'	WHITE, Edward	2°32'	SALINGER, Pierre
1°51'	CAGE, John	2°46'	ANNENBERG, Walter

OPPOSITION ♂ 4° orb

0°05'	SCHLESINGER, Arthur	2°02'	HAYES, Bob
0°10'	DONOVAN, Hedley	2°49'	PHILLIPS, Wendell
0°12'	WALTON, Bill	2°57'	ALIOTO, Joseph
1°03'	JACKSON, Henry	3°07'	HANSEN, Fred
1°49'	EAGLETON, Thomas	3°55'	SHORE, Dinah
1°50'	CROSSFIELD, Albert		

JUPITER/URANUS

CONJUNCTION ☌ 4° orb

0°10'	KITT, Eartha	1°46'	BLEGEN, Judith
0°38'	HAUGE, Gabriel	2°38'	MONDALE, Walter
0°42'	GIBSON, Althea	2°57'	O'NEAL, Ryan
1°28'	STOKES, Carl	3°02'	DYLAN, Bob
1°31'	VOLCKER, Paul	3°43'	Mc CLOSKEY, Paul

SEXTILE ⚹ 3° orb

0°23'	GOULIAN, Mehran	2°00'	DILLMAN, Bradford
0°45'	LAMBERT, Jack	2°08'	MUNSEL, Patrice
0°45'	CHILES, Lawton	2°15'	WALTON, Bill
0°56'	BOUTON, Jim	2°21'	RICE, Jim
1°07'	MURRAY, Don	2°24'	BAKER, Howard
1°48'	MAUCH, Gene	2°39'	LEDERBERG, Joshua
1°49'	BRADLEY, Bill	2°40'	TROWBRIDGE, Alexandre
2°00'	BERRA, Yogi	2°49'	CRAIN, Jeanne

SQUARE □ 3° orb

0°22'	BANKS, Ernie	1°12'	KERKORIAN, Kirk
0°31'	NOLAND, Kenneth	1°16'	MAYS, Willie
0°45'	JORDAN, William	1°51'	WHITE, Byron
0°53'	SLAYTON, Donald	1°58'	MORRISON, Toni
0°54'	BRANDO, Marlon	2°00'	BALL, William
1°01'	BYRNE, Brendan	2°24'	GROZA, Louis
1°09'	VAUGHAN, Sarah	2°31'	KASTENMEIER, Robert
1°10'	GOLD, Herbert	2°46'	BLUME, Judy

TRINE △ 3° orb

0°35'	NITSCHKE, Raymond	1°00'	WYETH, James
0°35'	KOCH, John	1°20'	TALESE, Gay
0°36'	BEBAN, Gary	1°35'	HAYES, Elvin
0°36'	MATHEWS, Edwin	1°45'	WILLSON, Zack
0°40'	BACKE, John	1°50'	GARDNER, Ava
0°42'	FOXX, Redd	2°50'	FINGERS, Rollie
0°48'	GODDARD, James	2°54'	MILLER, Neal
0°51'	GRIER, Rosey	2°57'	O'BRIEN, Parry
0°59'	GREGORY, Cynthia		

OPPOSITION ☍ 4° orb

0°10'	STEWART, William	0°55'	JONES, Everett
0°11'	BERRIGAN, Daniel	1°32'	STARR, Bryan
0°48'	AARON, Hank	2°04'	HILLS, Carla
0°54'	FREEMAN, David	2°34'	CHAMPION, Gower

SATURN/URANUS

CONJUNCTION ☌ 4° orb

1°25'	ROSE, Peter	3°43'	IRVING, John
1°50'	HARRELSON, Ken	3°46'	REED, Willis

SEXTILE ✳ 3° orb

0°33'	KESEY, Ken	1°51'	WRIGHT, Mickey
0°40'	LOGAN, Karen	1°57'	PLUNKETT, Jim
0°40'	MITCHELL, L.	2°03'	STEINEM, Gloria
0°57'	FOYT, Anthony	2°08'	PATTERSON, Floyd
1°13'	BAILAR, Benjamin	2°09'	MANN, Dick
1°21'	DELLUMS, Ronald	2°14'	RICHARDSON, Bobby
1°45'	PRESLEY, Elvis	2°23'	BLUE, Vida
1°48'	MOYERS, Bill	2°40'	CERNAN, Eugene

SQUARE □ 3° orb

0°10'	CHILES, Lawton	1°24'	NASH, Philleo
0°12'	DILLMAN, Bradford	1°44'	SNYDER, Gary
0°15'	MATHEWS, Edwin	2°18'	CASPER, William
1°12'	MATSON, Oliver	2°25'	ROBERTS, Ken
1°22'	RATHER, Dan	2°36'	WALTON, Bill

TRINE △ 3° orb

0°20'	NABER, John	1°40'	HEFNER, Hugh
0°25'	HALDEMAN, Harry	1°50'	NEAL, Patricia
0°26'	VOLCKER, Paul	1°50'	RICHARDS, Bob
0°27'	BERGQUIST, Kenneth	1°53'	COHEN, Wilbur
0°34'	LEWIS, Jerry	1°55'	HART, Phil
0°44'	STOKES, Carl	2°02'	HAUGE, Gabriel
1°12'	CURTIS, Ann	2°05'	TETLEY, Glen
1°15'	GIBSON, Althea	2°05'	ARFONS, Art
1°15'	KIRK, Claude	2°05'	Mc CLOSKEY, Paul
1°29'	ROZELLE, P.	2°06'	MURAYAMA, Makio
1°36'	GAGNE, Verne	2°39'	CHILD, Julia
1°40'	HERMAN, Woody	2°56'	BROWER, David

OPPOSITION ☍ 4° orb

2°09'	ASH, Roy	2°50'	CHAMBERLAIN, Owen
2°23'	DUERK, Alene	3°31'	GRAHAM, Billy
2°26'	LILLEHEI, Clarence	3°38'	CHADWICK, Florence
2°26'	DALY, James	3°45'	HARTZOG, George
2°47'	COSELL, Howard	3°55'	TALBERT, Bill

NEPTUNE/URANUS

CONJUNCTION ☌ 4° orb

<none>			

SEXTILE ⚹ 3° orb

<none>			

SQUARE □ 3° orb

0°03'	NABER, John	1°37'	GRIFFIN, Archie
0°08'	THOMPSON, D.	2°03'	BABASHOFF, Shirley
1°23'	FIDRYCH, Mark	2°51'	STONES, Dwight

TRINE △ 3° orb

0°04'	BUTKUS, Dick	1°21'	WELCH, Raquel
0°09'	BLACK, Karen	1°31'	THEROUX, Paul
0°12'	SMEAL, Eleanor	1°41'	DYLAN, Bob
0°12'	LILLY, Bob	1°51'	WHITWORTH, Kathy
0°14'	BLEGEN, Judith	1°54'	MORIARTY, Michael
0°20'	ROSE, Peter	2°07'	GABRIEL, Roman
0°25'	JOHNSON, Don	2°08'	ALWORTH, Lance
0°27'	HAYES, Bob	2°08'	SHEEN, Martin
0°27'	ASHLEY, Elisabeth	2°11'	KIDD, Billy
0°44'	KILMER, Bill	2°20'	IRVING, John
0°44'	WARFIELD, Paul	2°25'	OATES, J. C.
0°45'	O'NEAL, Ryan	2°28'	WILLIAMS, Billy
0°54'	FRAZIER, Joe	2°39'	ZIEGLER, Ronald
0°56'	OLSEN, Merlin	2°51'	ROSS, Diana
0°56'	DEBUSSCHERE, David	2°58'	BARRY, Richard
1°20'	DENVER, John	2°59'	FARROW, Mia

OPPOSITION ☍ 4° orb

0°02'	KROL, John	1°53'	NASH, Philleo
0°20'	MURPHY, William	1°56'	LOSEY, Joseph
0°22'	RAINEY, Froelich	2°01'	ROGERS, Roy
0°29'	KOCH, John	3°37'	GARDNER, John
0°30'	MILLER, Neal	3°47'	RUSH, David
0°55'	SIX, Robert		

PLUTO/URANUS

CONJUNCTION ☌ 4° orb

<none>	

SEXTILE ✶ 3° orb

0°04'	DAVENPORT, Willie	1°27'	LUCAS, George
0°09'	CONIGLIARO, Tony	1°31'	SEAVER, Tom
0°17'	FARROW, Mia	1°33'	RALSTON, R.
0°26'	LEVINE, James	1°34'	MAHAN, Larry
0°26'	THOMAS, Michael	1°36'	KING, Billie
0°39'	MORGAN, Joe	1°38'	Mc BRIDE, Patricia
0°48'	BARRY, Richard	1°42'	OWEN, Nancy
0°51'	ASHE, Arthur	2°05'	REED, Willis
0°52'	ROSS, Diana	2°29'	KAHN, Madeliine
0°53'	SAYERS, Gale	2°37'	DENVER, John
0°55'	MATSON, Randi	2°38'	PEPPLER, Mary
0°55'	GRAEBNER, Clark	2°39'	KIDD, Billy
0°59'	PERRINE, Valerie	2°44'	FRAZIER, Joe
1°07'	BRADLEY, Bill		

SQUARE □ 3° orb

0°09'	LAWRENCE, Carol	1°38'	BURKE, Yvonne
0°16'	STARR, Bryan	1°42'	BACKE, John
0°25'	HILLS, Carla	1°53'	MAYS, Willie
0°29'	GREY, Joel	2°06'	PEARSON, David
0°31'	BERRY, Raymond	2°07'	BLAKE, Robert
0°38'	ROTH, Philip	2°08'	LEWIS, Henry
0°49'	AARON, Hank	2°11'	BALL, William
0°55'	SIMONE, Nina	2°17'	PATTERSON, Floyd
0°56'	STEVENS, George	2°22'	PRESLEY, Elvis
0°56'	CASPER, William	2°35'	FOYT, Anthony
1°14'	BOWEN, William	2°43'	LEMON, Meadow
1°22'	NOVAK, Kim	2°50'	CERNAN, Eugene
1°33'	SCOTT, David	2°56'	NELSON, Willie
1°36'	GRIER, Rosey		

PLUTO/URANUS

TRINE △ 3° orb

0°04'	PAIGE, Janis	1°56'	STEWART, William
0°07'	GARDNER, Ava	1°57'	DOWNS, Hugh
0°12'	SCHROEDER, Frederick	2°00'	DUERK, Alene
0°17'	GILLIGAN, John	2°08'	Mc KISSICK, Floyd
0°18'	Mac NEIL, Cornell H.	2°10'	COSELL, Howard
0°19'	GLENN, John	2°11'	MAILER, Norman
0°20'	YOUNG, Whitney	2°13'	BENTSEN, Lloyd
0°24'	KRAMER, John	2°14'	CHAMBERLAIN, Owen
0°33'	HOLLEY, Robert	2°15'	Mc WHINNEY, Madeline
0°35'	MARSHALL, Burke	2°16'	KIRKLAND, Joseph
0°41'	ANDERSON, John	2°24'	VALENTI, Jack
0°47'	FOXX, Redd	2°28'	EISENHOWER, John
0°49'	CHARLES, Ezzard	2°35'	HARTZOG, George
0°53'	GALVIN, Robert	2°42'	FRIEDAN, Betty
0°54'	ANDERSON, Jack	2°54'	WILHELM, James
1°23'	CHAMPION, Gower	2°55'	CHANNING, Carol
1°53'	BERRIGAN, Daniel		

OPPOSITION ☍ 4° orb

<none>	

URANUS INGRESSES INTO SIGNS

1904 Dec 20	♑			2003 Mar 10	♓	
1912 Jan 30	♒			2003 Sep 15	♒	R
1912 Sep 4	♑	R		2003 Dec 30	♓	
1912 Nov 12	♒			2010 May 28	♈	
1919 Apr 1	♓			2010 Aug 14	♓	R
1919 Aug 16	♒	R		2011 Mar 12	♈	
1920 Jan 22	♓			2018 May 15	♉	
1927 Mar 31	♈			2018 Nov 6	♈	R
1927 Nov 4	♓	R		2019 Mar 6	♉	
1928 Jan 13	♈			2025 Jul 7	♊	
1934 Jun 6	♉			2025 Nov 8	♉	R
1934 Oct 10	♈	R		2026 Apr 26	♊	
1935 Mar 28	♉			2032 Aug 3	♋	
1941 Aug 7	♊			2032 Dec 12	♊	R
1941 Oct 5	♉	R		2033 May 22	♋	
1942 May 15	♊			2039 Aug 6	♌	
1948 Aug 30	♋			2040 Feb 25	♋	R
1948 Nov 12	♊	R		2040 May 15	♌	
1949 Jun 10	♋			2045 Oct 6	♍	
1955 Aug 24	♌			2046 Feb 8	♌	R
1956 Jan 28	♋	R		2046 Jul 22	♍	
1956 Jun 10	♌			2051 Dec 8	♎	
1961 Nov 1	♍			2052 Feb 1	♍	R
1962 Jan 10	♌	R		2052 Sep 11	♎	
1962 Aug 10	♍			2058 Nov 3	♏	
1968 Sep 28	♎			2059 Jun 1	♎	R
1969 May 20	♍	R		2059 Aug 11	♏	
1969 Jun 24	♎			2065 Jan 10	♐	
1974 Nov 21	♏			2065 May 1	♏	R
1975 May 1	♎	R		2065 Oct 28	♐	
1975 Sep 8	♏			2072 Jan 22	♑	
1981 Feb 17	♐			2072 Jun 25	♐	R
1981 Mar 20	♏	R		2072 Nov 12	♑	
1981 Nov 16	♐			2079 Mar 2	♒	
1988 Feb 15	♑			2079 Jul 13	♑	R
1988 May 27	♐	R		2079 Dec 23	♒	
1988 Dec 2	♑			2087 Feb 18	♓	
1995 Apr 1	♒			2094 Apr 28	♈	
1995 Jun 9	♑	R		2094 Sep 16	♓	R
1996 Jan 12	♒			2095 Feb 19	♈	

Index